A Practical Guide to Data Mining for Business and Industry

A Practical Guide to Data Mining for Business and Industry

Andrea Ahlemeyer-Stubbe
Director Strategic Analytics, DRAFTFCB München GmbH, Germany

Shirley Coleman
Principal Statistician, Industrial Statistics Research Unit
School of Maths and Statistics, Newcastle University, UK

Library of Congress Cataloging-in-Publication Data

Ahlemeyer-Stubbe, Andrea.
A practical guide to data mining for business and industry / Andrea Ahlemeyer-Stubbe,
Shirley Coleman.
     pages   cm
   Includes bibliographical references and index.
   ISBN 978-1-119-97713-1 (cloth)
1. Data mining.   2. Marketing–Data processing.   3. Management–Mathematical models.
I. Title.
   HF5415.125.A42 2014
   006.3′12–dc23
                                          2013047218

A catalogue record for this book is available from the British Library.

ISBN: 978-1-119-97713-1

Set in 10.5/13pt Minion by SPi Publisher Services, Pondicherry, India

Contents

Glossary of terms

Accuracy | A measurement of the match (degree of closeness) between predictions and real values.

Address | A unique identifier for a computer or site online, usually a URL for a website or marked with an @ for an email address. Literally, it is how your computer finds a location on the information highway.

Advertising | Paid form of a non-personal communication by industry, business firms, non-profit organisations or individuals delivered through the various media. Advertising is persuasive and informational and is designed to influence the purchasing behaviour and thought patterns of the audience. Advertising may be used in combination with sales promotions, personal selling tactics or publicity. This also includes promotion of a product, service or message by an identified sponsor using paid-for media.

Aggregation | Form of segmentation that assumes most consumers are alike.

Algorithm | The process a search engine applies to web pages so it can accurately produce a list of results based on a search term. Search engines regularly change their algorithms to improve the quality of the search results. Hence, search engine optimisation tends to require constant research and monitoring.

Analytics | A feature that allows you to understand (learn more) a wide range of activity related to your website, your online marketing activities and direct marketing activities. Using analytics provides you with information to help optimise your campaigns, ad groups and keywords, as well as your other online marketing activities, to best meet your business goals.

API | Application Programming Interface, often used to exchange data, for example, with social networks.

Attention | A momentary attraction to a stimulus, something someone senses via sight, sound, touch, smell or taste. Attention is the starting point of the perceptual process in that attention of a stimulus will either cause someone to decide to make sense of it or reject it.

B2B | Business To Business – Business conducted between companies rather than between a company and individual consumers. For example, a firm that makes parts that are sold directly to an automobile manufacturer.

B2C | Business To Consumer – Business conducted between companies and individual consumers rather than between two companies. A retailer such as Tesco or the greengrocer next door is an example of a B2C company.

Banner | Banners are the 468-by-60 pixels ad space on commercial websites that are usually 'hotlinked' to the advertiser's site.

Banner ad | Form of Internet promotion featuring information or special offers for products and services. These small space 'banners' are interactive: when clicked, they open another website where a sale can be finalized. The hosting website of the banner ad often earns money each time someone clicks on the banner ad.

Base period | Period of time applicable to the learning data.

Behavioural targeting | Practice of targeting and ads to groups of people who exhibit similarities not only in their location, gender or age but also in how they act and react in their online environment: tracking areas they frequently visit or subscribe to or subjects or content or shopping categories for which they have registered. Google uses behavioural targeting to direct ads to people based on the sites they have visited.

Benefit | A desirable attribute of goods or services, which customers perceive that they will get from purchasing and consuming or using them. Whereas vendors sell features ('a high-speed 1cm drill bit with tungsten-carbide tip'), buyers seek the benefit (a 1cm hole).

Bias | The expected value differs from the true value. Bias can occur when measurements are not calibrated properly or when subjective opinions are accepted without checking them.

Big data | Is a relative term used to describe data that is so large in terms of volume, variety of structure and velocity of capture that it cannot be stored and analysed using standard equipment.

Blog | A blog is an online journal or 'log' of any given subject. Blogs are easy to update, manage and syndicate, powered by individuals and/or corporations and enable users to comment on postings.

BOGOF | Buy One, Get One Free. Promotional practice where on the purchase of one item, another one is given free.

Boston matrix | A product portfolio evaluation tool developed by the Boston Consulting Group. The matrix categorises products into one of four classifications based on market growth and market share.

The four classifications are as follows:

- Cash cow – low growth, high market share
- Star – high growth, high market share
- Problem child – high growth, low market share
- Dog – low growth, low market share

Brand | A unique design, sign, symbol, words or a combination of these, employed in creating an image that identifies a product and differentiates or positions it from competitors. Over time, this image becomes associated with a level of credibility, quality and satisfaction in the consumers' minds. Thus, brands stand for certain benefits and value. Legal name for a brand is trademark, and when it identifies or represents a firm, it is called a brand name. (Also see Differentiation and Positioning.)

Bundling | Combining products as a package, often to introduce other products or services to the customer. For example, AT&T offers discounts for customers by combining 2 or more of the following services: cable television, home phone service, wireless phone service and Internet service.

Buttons | Objects that, when clicked once, cause something to happen.

Buying behaviour | The process that buyers go through when deciding whether or not to purchase goods or services. Buying behaviour can be influenced by a variety of external factors and motivations, including marketing activities.

Campaign | Defines the daily budget, language, geographic targeting and location of where the ads are displayed.

Cash cow | See 'Boston matrix'.

Category management | Products are grouped and managed by strategic business unit categories. These are defined by how consumers view goods rather than by how they look to the seller, for example, confectionery could be part of either a 'food' or 'gifts' category and marketed depending on the category into which it is grouped.

Channels | The methods used by a company to communicate and interact with its customers, like direct mail, telephone and email.

Characteristic | Distinguishing feature or attribute of an item, person or phenomenon that usually falls into either a physical, functional or operational category.

Churn rate | Rate of customers lost (stopped using the service) over a specific period of time, often over the course of a year. Used to compare against new customers gained.

Click | The opportunity for a visitor to be transferred to a location by clicking on an ad, as recorded by the server.

Clusters | Customer profiles based on lifestyle, demographic, shopping behaviour or appetite for fashion. For example, ready-to-eat meals may be heavily influenced by the ethnic make-up of a store's shoppers, while beer, wine and spirits categories in the same store may be influenced predominantly by the shopper's income level and education.

Code | Anything written in a language intended for computers to interpret.

Competitions | Sales promotions that allow the consumer the possibility of winning a prize.

Competitors | Companies that sell products or services in the same marketplace as one another.

Consumer | A purchaser of goods or services at retail, or an end user not necessarily a purchaser, in the distribution chain of goods or services (gift recipient).

Contextual advertising | Advertising that is targeted to a web page based on the page's content, keywords or category. Ads in most content networks are targeted contextually.

Cookie | A file on your computer that records information such as where you have been on the World Wide Web. The browser stores this information which allows a site to remember the browser in future transactions or requests. Since the web's protocol has no way to remember requests, cookies read and record a user's browser type and IP address and store this information on the user's own computer. The cookie can be read only by a server in the domain that stored it. Visitors can accept or deny cookies by changing a setting in their browser preferences.

Coupon | A ticket that can be exchanged for a discount or rebate when procuring an item.

CRM | Customer Relationship Management – Broad term that covers concepts used by companies to manage their relationships with customers, including the capture, storage and analysis of customer, vendor, partner and internal process information. CRM is the coherent management of contacts and interactions with customers. This term is often used as if it related purely to the use of Information Technology (IT), but IT should in fact be regarded as a facilitator of CRM.

Cross-selling | A process to offer and sell additional products or services to an existing customer.

Customer |A person or company who purchases goods or services (not necessarily the end consumer).

Customer Lifetime Value (CLV) | The profitability of customers during the lifetime of the relationship, as opposed to profitability on one transaction.

Customer loyalty | Feelings or attitudes that incline a customer either to return to a company, shop or outlet to purchase there again or else to repurchase a particular product, service or brand.

Customer profile | Description of a customer group or type of customer based on various geographic, demographic, and psychographic characteristics; also called shopper profile (may include income, occupation, level of education, age, gender, hobbies or area of residence). Profiles provide knowledge needed to select the best prospect lists and to enable advertisers to select the best media

Data | Facts/figures pertinent to customer, consumer behaviour, marketing and sales activities.

Data processing | The obtaining, recording and holding of information which can then be retrieved, used, disseminated or erased. The term tends to be used in connection with computer systems and today is often used interchangeably with 'information technology'.

Database marketing | Whereby customer information, stored in an electronic database, is utilised for targeting marketing activities. Information can be a mixture of what is gleaned from previous interactions with the customer and what is available from outside sources. (Also see 'Customer Relationship Management (CRM)'.)

Demographics | Consumer statistics regarding socio-economic factors, including gender, age, race, religion, nationality, education, income, occupation and family size. Each demographic category is broken down according to its characteristics by the various research companies.

Description | A short piece of descriptive text to describe a web page or website. With most search engines, they gain this information primarily from the metadata element of a web page. Directories approve or edit the description based on the submission that is made for a particular URL.

Differentiation | Ensuring that products and services have a unique element to allow them to stand out from the rest.

Digital marketing | Use of Internet-connected devices to engage customers with online products and service marketing/promotional programmes. It includes marketing mobile phones, iPads and other Wi-Fi devices.

Direct marketing | All activities which make it possible to offer goods or services or to transmit other messages to a segment of the population by post, telephone, email or other direct means.

Distribution | Movement of goods and services through the distribution channel to the final customer, consumer or end user, with the movement of payment (transactions) in the opposite direction back to the original producer or supplier.

Dog | See 'Boston matrix'.

Domain | A domain is the main subdivision of Internet addresses and the last three letters after the final dot, and it tells you what kind of organisation you are dealing with. There are six top-level domains widely used: .com (commercial), .edu (educational), .net (network operations), .gov (US government), .mil (US military) and .org (organisation). Other two-letter domains represent countries: .uk for the United Kingdom, .dk for Denmark, .fr for France, .de for Germany, .es for Spain, .it for Italy and so on.

Domain knowledge | General knowledge about in-depth business issues in specific industries that is necessary to understand idiosyncrasies in the data.

ENBIS | European Network of Business and Industrial Statistics.

ERP | | Enterprise Resource Planning includes all the processes around billing, logistics and real business processes.

ETL | Extraction, Transforming and Loading processes which cover all processes and algorithms that are necessary to take data from the original source to the data warehouse.

Forecast | The use of experience and/or existing data to learn/develop models that will be used to make judgments about future events and potential results. Often used interchangeably with prediction.

Forms | The pages in most browsers that accept information in text-entry fields. They can be customised to receive company sales data and orders, expense reports or other information. They can also be used to communicate.

Freeware | Shareware, or software, that can be downloaded off the Internet – for free.

Front-end applications | Interfaces and applications mainly used in customer service and help desks, especially for contacts with prospects and new customers.

ID | Unique identity code for cases or customers used internally in a database.

Index | The database of a search engine or directory.

Input or explanatory variable | Information used to carry out prediction and forecasting. In a regression, these are the X variables.

Inventory | The number of ads available for sale on a website. Ad inventory is determined by the number of ads on a page, the number of pages containing ad space and the number of page requests.

Key Success Factors (KSF) and Key Performance Indicators (KPIs) | Those factors that are a necessary condition for success in a given market. That is, a company that does poorly on one of the factors critical to success in its market is certain to fail.

Knowledge | A customer's understanding or relationship with a notion or idea. This applies to facts or ideas acquired by study, investigation, observation or experience, not assumptions or opinions.

Knowledge Management (KM) | The collection, organisation and distribution of information in a form that lends itself to practical application. Knowledge management often relies on IT to facilitate the storage and retrieval of information.

Log or log files | File that keeps track of network connections. These text files have the ability to record the amount of search engine referrals that is being delivered to your website.

Login | The identification or name used to access – log into – a computer, network or site.

Logistics | Process of planning, implementing and controlling the efficient and effective flow and storage of goods, services and related information from point of origin to point of consumption for the purpose of conforming to customer requirements, internal and external movements and return of materials for environmental purposes.

Mailing list | Online, a mailing list is an automatically distributed email message on a particular topic going to certain individuals. You can subscribe or unsubscribe to a mailing list by sending a message via email. There are many good professional mailing lists, and you should find the ones that concern your business.

Market research | Process of making investigations into the characteristics of given markets, for example, location, size, growth potential and observed attitudes.

Marketing | Marketing is the management process responsible for identifying, anticipating and satisfying customer requirements profitably.

Marketing dashboard | Any information used or required to support marketing decisions – often drawn from a computerised 'marketing information system'.

Needs | Basic forces that motivate a person to think about and do something/take action. In marketing, they help explain the benefit or satisfaction derived from a product or service, generally falling into the physical (air > water > food > sleep > sex > safety/security) or psychological (belonging > esteem > self-actualisation > synergy) subsets of Maslow's hierarchy of needs.

Null hypothesis | A proposal that is to be tested and that represents the baseline state, for example, that gender does not affect affinity to buy.

OLAP | Online Analytical Processing which is a convenient and fast way to look at business-related results or to monitor KPIs. Similar words are Management Information Systems (MIS) and Decision Support Systems (DSS).

Outlier | Outliers are unusual values that show up as very different to other values in the dataset.

Personal data | Data related to a living individual who can be identified from the information; includes any expression of opinion about the individual.

Population | All the customers or cases for which the analysis is relevant. In some situations, the population from which the learning sample is taken may necessarily differ from the population that the analysis is intended for because of changes in environment, circumstances, etc.

Precision | A measurement of the match (degree of uncertainty) between predictions and real values.

Prediction | Uses statistical models (learnt on existing data) to make assumptions about future behaviour, preferences and affinity. Prediction modelling is a main part of data mining. Often used interchangeably with forecast.

Primary key | A primary key is a field in a table in a database. Primary keys must contain unique, non-null values. If a table has a primary key defined on any field(s), then you cannot have two records having the same value of that field(s).

Probability | The chance of something happening.

Problem child | See 'Boston matrix'.

Product | Whatever the customer thinks, feels or expects from an item or idea. From a 'marketing-oriented' perspective, products should be defined by what they satisfy, contribute or deliver versus what they do or the form utility involved in their development. For example, a dishwasher cleans dishes but it's what the consumer does with the time savings that matters most. And ultimately, a dishwasher is about 'clean dishes', not the act of cleaning them.

Prospects | People who are likely to become users or customers.

Real Time | Events that happen in real time are happening virtually at that particular moment. When you chat in a chat room or send an instant message, you are interacting in real time since it is immediate.

Recession | A period of negative economic growth. Common criteria used to define when a country is in a recession are two successive quarters of falling GDP or a year-on-year fall in GDP.

Reliability | Research study can be replicated and get some basic results (free of errors).

Re-targeting | Tracking website visitors, often with small embedded coding on the visitor's computer called 'cookies'. Then displaying relevant banner ads relating to products and services on websites previously visiting as surfers visit other websites.

Return On Investment (ROI) | The value that an organisation derives from investing in a project. Return on investment = (revenue – cost)/cost, expressed as a percentage. A term describing the calculation of the financial return on an Internet marketing or advertising initiative that incurs some cost. Determining the ROI and the actual ROI in Internet marketing and advertising has been much more accurate than television, radio and traditional media.

Revenue | Amounts generated from sale of goods or services, or any other use of capital or assets before any costs or expenses are deducted. Also called sales.

RFM | A tool used to identify best and worst customers by measuring three quantitative factors:

- Recency – How recently a customer has made a purchase
- Frequency – How often a customer makes a purchase
- Monetary value – How much money a customer spends on purchases

RFM analysis often supports the marketing adage that '80% of business comes from 20% of the customers'. RFM is widely used to split customers into different segments and is an easy tool to predict who will buy next.

Sample and sampling | A sample is a statistically representative subset often used as a proxy for an entire population. The process of selecting a suitable sample is referred to as sampling. There are different methods of sampling including stratified and cluster sampling.

Scorecard | Traditionally, a scorecard is a rule-based method to split subjects into different segments. In marketing, a scorecard is sometimes used as an equivalent name for a predictive model.

Segmentation | Clusters of people with similar needs that share other geographic, demographic and psychographic characteristics, such as veterans, senior citizens or teens.

Session | A series of transactions or hits made by a single user. If there has been no activity for a period of time, followed by the resumption of activity by the same user, a new session is considered started. Thirty minutes is the most common time period used to measure a session length.

Significance | An important result; statistical significance means that the probability of being wrong is small. Typical levels of significance are 1%, 5% and 10%.

SQL | Standard Query Language, a programming language to deal with databases.

Star | See 'Boston matrix'.

Supervised learning | Model building when there is a target and information is available that can be used to predict the target.

Tags | Individual keywords or phrases for organising content.

Targeting | The use of 'market segmentation' to select and address a key group of potential purchasers.

Testing (statistical) | Using evidence to assess the truth of a hypothesis.

Type I error | Probability of rejecting the null hypothesis when it is true, for example, a court of law finds a person guilty when they are really innocent.

Type II error | Probability of accepting the null hypothesis when it is false, for example, a court of law finds a person innocent when they are really guilty.

Unsupervised learning | Model building when there is no target, but information is available that can describe the situation.

URL | Uniform resource locator used for web pages and many other applications.

Validity | In research studies, it means the data collected reflects what it was designed to measure. Often, invalid data also contains bias.

X variable | Explanatory variable used in a data mining model.

Y variable | Dependent variable used in a data mining model also called target variable.

PART I

DATA MINING CONCEPT

1

Introduction

Introduction

1.1 AIMS OF THE BOOK

The power of data mining is a revelation to most companies. Data mining means extracting information from meaningful data derived from the mass of figures generated every moment in every part of our life. Working with data every day, we realise the satisfaction of unearthing patterns and meaning. This book is the result of detailed study of data and showcases the lessons learnt

---

A Practical Guide to Data Mining for Business and Industry, First Edition.
Andrea Ahlemeyer-Stubbe and Shirley Coleman.
© 2014 John Wiley & Sons, Ltd. Published 2014 by John Wiley & Sons, Ltd.
Companion website: www.wiley.com/go/data_mining

when dealing with data and using it to make things better. There are many tricks of the trade that help to ensure effective results. The statistical analysis involved in data mining has features that differentiate it from other types of statistics. These insights are presented in conjunction with background information in the context of typical scenarios where data mining can lead to important benefits in any business or industrial process.

A Practical Guide to Data Mining for Business and Industry:

- Is built on expertise from running consulting businesses.
- Is written in a practical style that aims to give tried and tested guidance to finding workable solutions to typical business problems.
- Offers solution patterns for common business problems that can be adapted by the reader to their particular area of interest.
- Has its focus on practical solutions, but the book is grounded on sound statistical practice.
- Is in the style of a cookbook or blueprint for success.

Inside the book, we address typical marketing and sales problems such as 'finding the top 10% of customers likely to buy a special product'. The content focuses on sales and marketing because domain knowledge is a major part of successful data mining and everybody has the domain knowledge needed for these types of problems. Readers are unlikely to have specific domain knowledge in other sectors, and this would impair their appreciation of the techniques. We are all targeted as consumers and customers; therefore, we can all relate to problems in sales and marketing. However, the techniques discussed in the book can be applied in any sector where there is a high volume of observed but possibly 'dirty' data in need of analysis. In this scenario, statistical analysis appropriate to data from designed experiments cannot be used. To help in adapting the techniques, we also consider examples in banking and insurance. Finally, we include suggestions on how the techniques can be transferred to other sectors.

The book is distinctly different from other data mining books as it focuses on finding smart solutions rather than studying smart methods. For the reader, the book has two distinct benefits: on the one hand, it provides a sound foundation to data mining and its applications, and on the other hand, it gives guidance in using the right data mining method and data treatment.

The overall goal of the book is to show how to make an impact through practical data mining.

Some statistical concepts are necessary when data mining, and they are described in later chapters. It is not the aim of the book to be a statistical textbook. The Glossary covers some statistical terms, and interested readers should have a look at the Bibliography.

The book is aimed at people working in companies or other people wanting to use data mining to make the best of their data or to solve specific practical problems. It is suitable for beginners in the field and also those who want to expand their knowledge of handling data and extracting information. A collection of standard problems is addressed in the recipes, and the solutions proposed are those using the most efficient methods that will answer the underlying business question. We focus on methods that are widely available so that the reader can readily get started.

1.2 DATA MINING CONTEXT

Modern management is data driven; customers and corporate data are becoming recognised as strategic assets. Decisions based on objective measurements are better than decisions based on subjective opinions which may be misleading and biased. Data is collected from all sorts of input devices and must be analysed, processed and converted into information that informs, instructs, answers or otherwise aids understanding and decision making. Input devices include cashier machines, tills, data loggers, warehouse audits and Enterprise Resource Planning (ERP) systems. The ability to extract useful but usually hidden knowledge from data is becoming increasingly important in today's competitive world. When the data is used for prediction, future behaviour of the business is less uncertain and that can only be an advantage; 'forewarned is forearmed'!

As Figure 1.1 shows, the valuable resource of historical data can lead to a predictive model and a way to decide on accepting new applicants to a business scheme.

Data mining solution

Utilise data from the past (historical data of an organisation)
to predict activities on future applicants

FIGURE 1.1   Data mining short process.

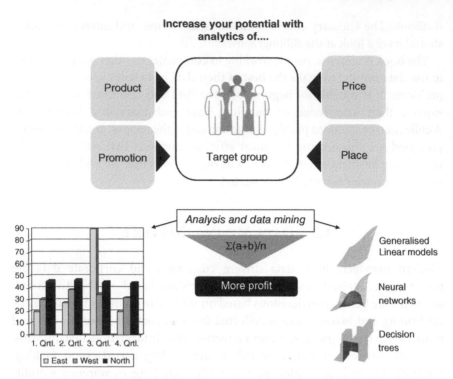

FIGURE 1.2 Increasing profit with data mining.

With technological advancements, the computer industry has witnessed a tremendous growth in both hardware and software sectors. Sophisticated databases have encouraged the storage of massive datasets, and this has opened up the need for data mining in a range of business contexts. Data mining, with its roots in statistics and machine learning, concerns data collection, description, analysis and prediction. It is useful for decision making, when all the facts or data cannot be collected or are unknown. Today, people are interested in knowledge discovery (i.e. intelligence) and must make sense of the terabytes of data residing in their databases and glean the important patterns from it with trustworthy tools and methods, when humans can no longer juggle all these data and analyses in their heads (see Figure 1.2).

1.2.1 Domain Knowledge

We will refer to the concept of domain knowledge very often in the text to follow. Domain knowledge is all the additional information that we have about a situation; for example, there may be gaps in the data, and our domain knowledge may be able to tell us that the sales process or production was halted for that period. We can now treat the data accordingly as it is not really zero, or missing in the sense of being omitted, but is zero for a distinct reason. Domain knowledge includes meta-data. For example, we may be monitoring sales of a product, and our main interest is in the quantities sold and their sale price. However, meta-data about the level of staffing in the sales outlet may also give us information to help in the interpretation.

1.2.2 Words to Remember

The results of an analysis are referred to in different ways. The model itself can also be referred to as a scorecard for the analysis. Each customer will have their own score based on the scorecard that was implemented. For example, a customer may have a score for their affinity to buy a cup of coffee, and there will be a scorecard indicating the structure of the model predicting the affinity. The term scorecard comes from earlier days when models were simpler, and typically, a customer collected a score when they carried out a certain behaviour. An example of this type of modelling is the Recency, Frequency and Monetary Value (RFM) method of segmentation in which the scores are given for the customer's RFM and the scores are combined together to identify high- and low-worth customers.

1.2.3 Associated Concepts

A lot of Customer Relations Management (CRM) analysis is complementary to information on the company reports and Marketing Dashboard (MD). For example, the MD may typically contain a summary of purchases of customers in different groupings and how they have changed from previous quarters or years. The numbers may be actual or predicted or a combination of the two.

The customer grouping results can be those who buy in the summer, for example, or those who have a response rate of 20%; the grouping could be for a particular campaign or averaged over a wider period.

Key Performance Indicators (KPIs) are a group of measurements and numbers that help to control the business and can be defined in detail down to the campaign level and for special marketing activities. Typical examples for KPIs are click rate, response rate, churn rate and cost per order. They are a convenient way to present overall performance in a succinct manner although care has to be taken that important details are not overlooked.

Analytics is the general name for data analysis and decision making. Descriptive analytics focuses on describing the features of data, and predictive analytics refers to modelling.

1.3   Global Appeal

In the business world, methods of communicating with customers are constantly changing. In this book, we direct most of our attention to businesses that have direct communication with customers. Direct communication means that the company actively promotes their products. Promotion can be through email contact, brochures, sales representatives, web pages and social media.

Whatever the means of contact, companies are increasingly becoming aware that their vast reserves of data contain a wealth of information. Large companies such as supermarkets and retail giants have been exploiting this source of information for many years, but now, smaller businesses are also becoming aware of the possibilities. Apart from marketing and advertising, production and finance are also benefitting from data mining. These sectors use the same methods and mechanisms as marketing and advertising; however, we have tended to use marketing data to illustrate the methods because it is easier to relate to and does not require specific technical details about the product or knowledge about the production process; everyone is familiar with sales because we are all part of the target audience and we are all affected by the results of the data mining carried out by large companies.

Institutions like healthcare establishments and government are also tapping into their data banks and finding that they can improve their services and increase their efficiency by analysing their data in a focused way.

Making use of data requires a scientific approach and a certain amount of technical skill. However, people working in all types of company are now becoming more adept with data manipulation; the techniques and recipes described in this book are accessible to all businesses, large and small.

1.4   Example Datasets Used in This Book

Although there are many different datasets, they all share common characteristics in terms of a required output and explanatory input. For illustrative purposes, the pre-analytics and analytics described in Part II of this book are applied to typical datasets.

One dataset is from a mail-order warehouse; this is chosen because it is a familiar concept to everyone even if the application for your data mining is quality engineering, health, finance or any other area. The dataset includes purchase details, communication information and demographics and is a subset of a large real dataset used for a major data mining exercise. There are 50 000 customers that are a sample from the full dataset, and you will see in the ensuing steps how the dataset is put into shape for effective data mining (see Figure 1.3).

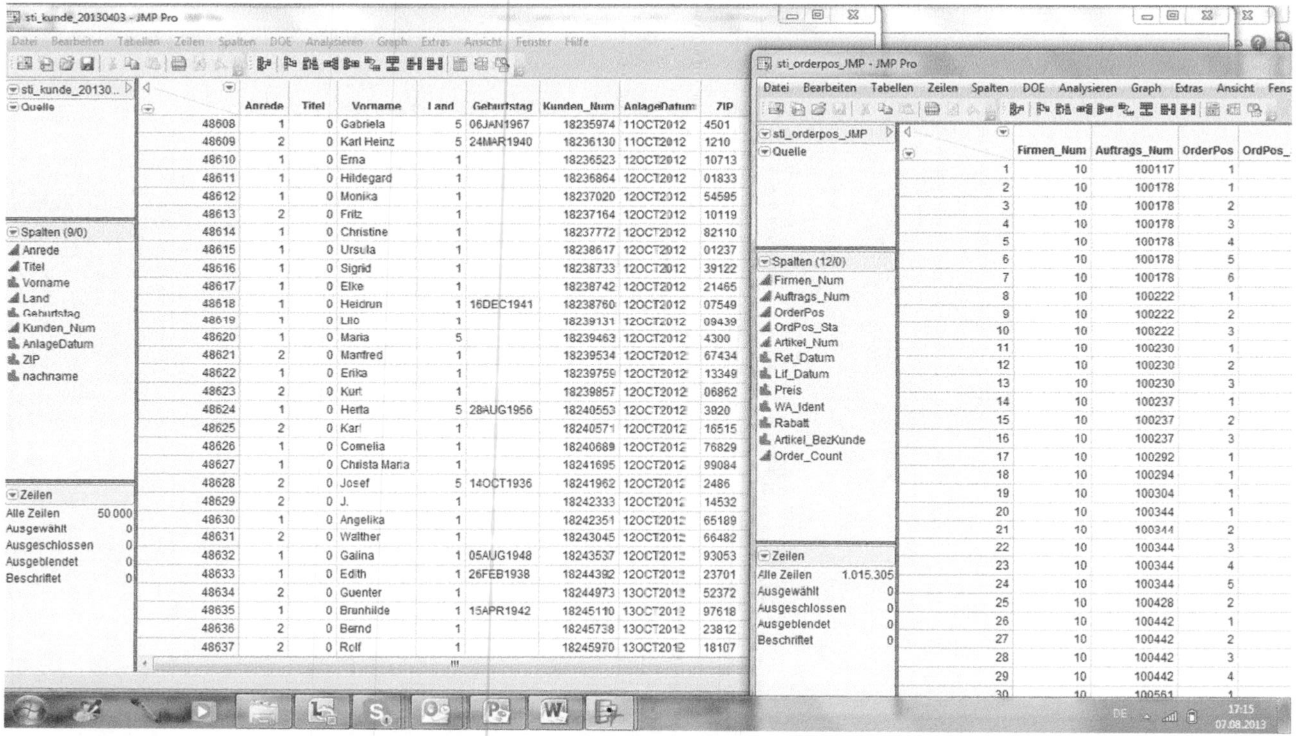

FIGURE 1.3   Example data – 50 000 sample customers and table of order details.

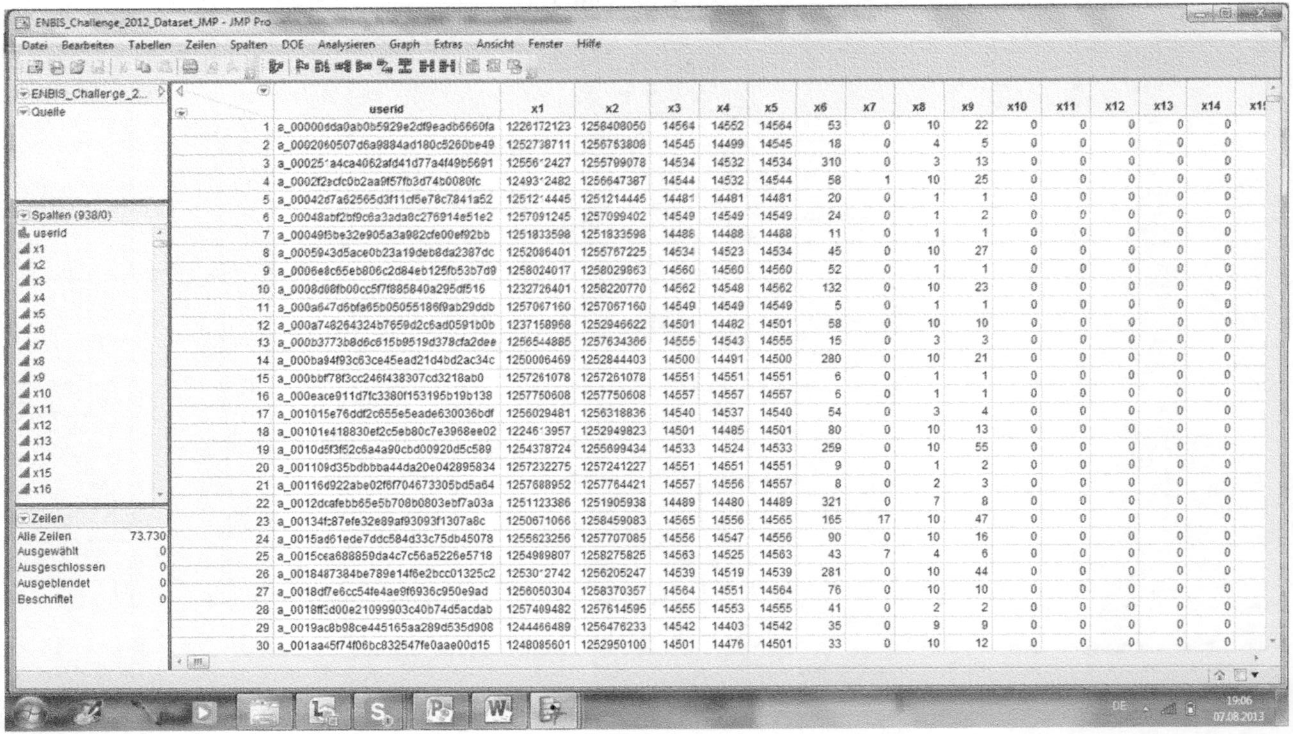

FIGURE 1.4 Example data – ENBIS Challenge.

Initially, there are around 200 variables, but these will be augmented as described in the following text.

Another dataset is from web analytics. This was the basis of the European Network of Business and Industrial Statistics (ENBIS) Challenge in 2012 (see Figure 1.4).

Most of the calculations in this book have been carried out using JMP software or tools from the SAS analytical software suite. JMP and SAS are well-established analytical software, and there are many others available. The guidelines for choosing software are given in Chapter 12.

1.5   RECIPE STRUCTURE

A cookbook should have easy-to-follow instructions, and our aim is to show how data mining methods can be applied in practice. Data mining analytical methods have subtle differences from statistical analysis, and these have been highlighted in the text along with the guidelines for data preparation and methods.

There are standard analyses that are required over and over again, and Part III of the book gives details of these. The recipes are grouped in four parts: prediction, intra-customer analysis, learning from a small dataset and miscellaneous recipes. Each of the generic recipes is described in full, and within each of them, there are modifications which are added as adaptations.

The full recipe structure is given in detail below. Not all of the components are included for each recipe, and the adaptations just have the components which make them differ from the generic recipe.

Industry: This refers to the area or sector of applications, for example, mail-order businesses, publishers, online shops, department stores or supermarkets (with loyalty cards) or everybody using direct communication to improve business.

Areas of interest: This is specific, for example, marketing, sales and online promotions.

Challenge: This could be, for example, to find and address the right number of customers to optimise the return on investment of a marketing campaign.

Typical application: This is more specific, for example, to prepare for summer sales promotions.

Necessary data: This is all the data that is vital for the analysis. The data must have some direct relationship to the customer reactions or must have come directly from the customer (e.g. data directly from the purchasing process or marketing activities).

Population: This is defined according to the problem and the business briefing. Note that campaigns can be highly seasonal in which case we need to consider the population for at least one cycle.

Target variable: This is the decisive variable of interest, for example, a binary variable such as 'buying' or 'not buying', or it could be a metric-level quantity like number or value of sales.

Input data – must-haves: These are the key variables upon which the analysis depends.

Input data – nice to haves: These are other variables that could improve the modelling but may be more difficult to find or to construct.

Data mining methods: There are often a few different methods that could be used, and these are listed here.

How to do it: The sections from Data preparation to Implementation give details of what to do.

Data preparation: The specific features of preparing data for each recipe are described here.

Business issues: These may include strategy changes involving, for example, sales channels, locations, diversity or products. These considerations should be borne in mind when analysing the data.

Transformation: For example, the target and/or input variables may need to be classified or converted to indicator variables. Other variables may require transformations to ameliorate asymmetries.

Marketing database: This refers to creating the dataset from which the analysis can be conducted.

Analytics: The sections from Partitioning to Validation are the step-by-step account of the analysis.

Partitioning the data: This may include consideration of sample size, stratification and other issues.

Pre-analytics: This describes the work needed prior to analysis. It may involve screening out some variables, for example, variables that have zero value or are all one value. Feature selection can also be done at this stage.

Model building: Models are built by obtaining the best-fit formulae or membership rules, for example, in cluster analysis.

Evaluation: Evaluation focuses on how well the analytical process has performed in terms of its value to the business. It also considers the quality of the model as regards its usefulness for decision making. Model validation is an important aspect of evaluation, and so these two are often considered together.

Validation: Validation focuses on making sure that the solution addresses the business problem. It may utilise face validation which involves comparison of the common viewpoint with the results of the modelling. It also considers how well the model fits the data. This usually involves applying the model to different subsets of the data and comparing the results.

Implementation: Here, we address the original statement of the recipe, such as how to name and address the right number of customers, and discuss how the model can be put into practice.

Hints and tips: These are specific to the particular recipe and may include suggestions for refreshing the models.

How to sell to management: This is a very important part and includes tables and plots that may make the results catchy and appealing.

1.6 FURTHER READING AND RESOURCES

There is an enthusiastic constituency of data miners and data analysts. Besides creating informative websites and meeting at conferences, they have developed some interesting communal activities like various challenges and competitions. One long-running competition is the Knowledge Discovery and Data Mining (KDD) Cup in the United States. The KDD website provides a wealth of interesting datasets and solutions to challenge questions offered by competitors.

The DATA-MINING-CUP in Germany is aimed mostly at students. There is also the ENBIS Challenge. In 2012, the challenge was around an enormous set of clickstream data produced when users clicked through web pages of a particular company. The challenge was to identify groups of people for whom the company could tailor promotional attention. In 2010 and 2011, the challenge was focused around some pharmaceutical data, and in 2009, a vast set of sales data was made available with the challenge of identifying patterns of behaviour. More information about ENBIS and the ENBIS Challenges can be found at www.enbis.org.

In addition to these resources, there are many community websites, annual conferences and games available.

2

Data Mining Definition

Data Mining Definition

A Practical Guide to Data Mining for Business and Industry, First Edition.
Andrea Ahlemeyer-Stubbe and Shirley Coleman.
© 2014 John Wiley & Sons, Ltd. Published 2014 by John Wiley & Sons, Ltd.
Companion website: www.wiley.com/go/data_mining

2.1   Types of Data Mining Questions

Data mining covers a wide range of activities. It seeks to provide the answer to questions such as these:

- What is contained in the data?
- What kinds of patterns can be discerned from the maze of data?
- How can all these data be used for future benefit?

2.1.1   Population and Sample

In data mining, datasets can be enormous – there may be millions of cases. Different types of industry, however, vary a lot as regards the number of cases emerging from the business processes. Web applications, for example, may collect data from millions of cookies, whereas other applications, like loyalty clubs or CRM programmes, may have more limited cases. Data protection laws and local market and industry customs vary, but in many countries, it is possible to purchase or to rent information at both a detailed and a summary or aggregate level.

Data mining uses the scientific method of exploration and application. We are presented with a mass of data that in some cases we can consider as a whole population. In other words, we have all the information that there is. In other cases, our dataset may be considered as a large sample. If we are dealing with smallish amounts of data (up to 10 000 cases), then we may prefer to work with the whole dataset. If we are dealing with larger datasets, we may choose to work with a subset for ease of manipulation. If the analysis is carried out on a sample, the implication is that the results will be representative of the whole population. In other words, the results of the analysis on the sample can be generalised to be relevant for the whole population.

The sample therefore has to be good, by which we mean that it has to be representative and unbiased. Sampling is a whole subject in itself. As we are usually dealing with large populations and can afford to take large samples, we can take a random sample in which all members of the population have an equal chance of being selected. We will revisit the practical issues around sampling in other sections of the book. We may also partition the dataset into several samples so that we can test our results. If we have a small dataset, then we resample by taking random subsets within the same sample, referred to as bootstrapping. We then have to consider ways of checking that the resulting sample is representative.

Sometimes, we only consider a part of the population for a particular analysis, for example, we may only be interested in buying behaviour around Christmas or in the summer months. In this case, the subset is referred to as a sampling frame as it is just from this subset that further samples will be selected.

2.1.2   Data Preparation

Data preparation for data mining is a vital step that is sometimes overlooked. From our earliest years, we have been taught that 'two plus two equals four'. Numbers are seen as concrete, tangible, solid, inevitable, beyond argument and a tool that can be used to measure anything and everything. But numbers have inherent variation, for example, two products may have been sold on a certain day, but their sale price may be different; interpretations made at face value may not be true. Some businesses use data for decision making without even making sure that the data is meaningful, without first transforming the data into knowledge and finally into intelligence. 'Intelligence' comes from data which has been verified for its validity through the use of past experience and has been described from considerations of its context.

2.1.3   Supervised and Unsupervised Methods

Data mining is a process that uses a variety of data analysis methods to discover the *unknown, unexpected, interesting and relevant* patterns and relationships in data that may be used to make valid and accurate predictions. In general, there are two methods of data analysis: supervised and unsupervised (see Figure 2.1 and Figure 2.2). In both cases, a sample of observed data is required. This data

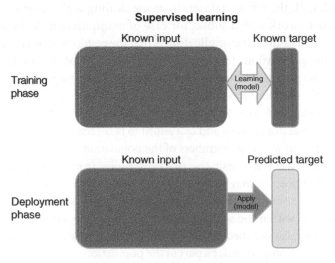

Figure 2.1   Supervised learning.

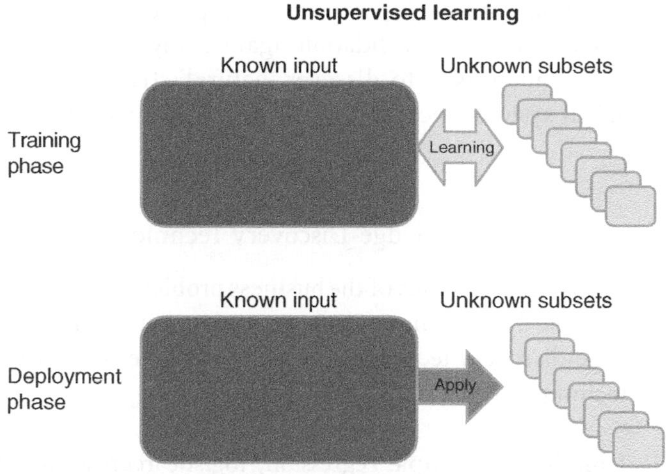

Unsupervised learning

FIGURE 2.2   Unsupervised learning.

may be termed the training sample. The training sample is used by the data mining activities to learn the patterns in the data.

Supervised data analysis is used to estimate an unknown dependency from known input output data. Input variables might include the quantities of different articles bought by a particular customer, the date they made the purchase, the location and the price they paid. Output variables might include an indication of whether the customer responds to a sales campaign or not. Output variables are also known as targets in data mining. In the supervised environment, sample input variables are passed through a learning system, and the subsequent output from the learning system is compared with the output from the sample. In other words, we try to predict who will respond to a sales campaign. The difference between the learning system output and the sample output can be thought of as an error signal. Error signals are used to adjust the learning system. This process is done many times with the data from the sample, and the learning system is adjusted until the output meets a minimal error threshold. It is the same process taken to fine-tune a newly bought piano. The fine-tuning could be done by an expert or by using some electronic instrument. The expert provides notes for the training sample, and the newly bought piano is the learning system. The tune is perfected when the vibration from the keynotes of the piano matches the vibration in the ear of the expert.

Unsupervised data analysis does not involve any fine-tuning. Data mining algorithms search through the data to discover patterns, and there is no target or aim variable. Only input values are presented to the learning system without the need for validation against any output. The goal of unsupervised data analysis is to discover 'natural' structures in the input data. In biological systems, perception is a task learnt via an unsupervised technique.

2.1.4 Knowledge-Discovery Techniques

Depending on the characteristics of the business problems and the availability of 'clean' and suitable data for the analysis, an analyst must make a decision on which knowledge-discovery techniques to use to yield the best output. Among the available techniques are:

- **Statistical methods**: multiple regression, logistic regression, analysis of variance and log-linear models and Bayesian inference
- **Decision trees and decision rules**: Classification And Regression Tree (CART) algorithms and pruning algorithms
- **Cluster analysis**: divisible algorithm, agglomerative algorithms, hierarchical clustering, partitional clustering and incremental clustering
- **Association rules**: market basket analysis, *a priori* algorithm and sequence patterns and social network analysis
- **Artificial neural networks**: multilayer perceptrons with back-propagation learning, radial networks, Self-Organising Maps (SOM) and Kohonen networks
- **Genetic algorithms**: used as a methodology for solving hard optimisation problems
- **Fuzzy inference systems**: based on theory of fuzzy sets and fuzzy logics
- **N-dimensional visualisation methods**: geometric, icon-based, pixel-oriented and hierarchical techniques
- **Case-Based Reasoning (CBR)**: based on comparing new cases with stored cases, uses similarity measurements and can be used when only a few cases are available

This list is not exhaustive, and the order does not suggest any priority in the application of these techniques. This book will concentrate on the widely used methods that are implemented in a wide range of data mining software products and those methods that are known to deliver good results on business questions in a relatively short time. We will focus more on the business need than on the scientific aspects. The Bibliography contains references to literature that covers all of these techniques.

2.2   DATA MINING PROCESS

The need for data mining arises from the realisation that improvements can be made to the way a business is run. The logical first step is to understand business needs and identify and prioritise areas needing attention. These are typically: too many dropout customers, disappointing sales, geographic areas with unnecessarily poor returns or quality issues or, on the more positive side, how to turn potential customers into customers or develop areas with opportunities. Many of these questions can be tackled by looking at relevant data.

All data mining analytics should follow a defined process that ensures the quality of the results. There are different data mining process definitions available which are similar in essence, for example, CRISP-DM and SEMMA.

In general, the data mining process is shown in Figure 2.3.

Regardless of the area of application or the concrete problem definition, the theoretical process can be mapped by the following steps:

1. Business task: clarification of the business question behind the problem
2. Data: provision and processing of the required data
3. Modelling: analysis of the data
4. Evaluation and validation during the analysis stage
5. Application of data mining results and learning from the experience

These steps are an example of a business improvement or audit cycle. Each will now be discussed in further detail.

FIGURE 2.3   The general data mining process.

2.3 Business Task: Clarification of the Business Question behind the Problem

As with all scientific endeavours, it is most important to be clear about what you are trying to do. Here, we consider the problem definition and specification of the aim of the action, the planned application and the application period. Consider the scenario of the data miner discussing a project with a customer or client.

A vital part of the problem definition is to have a written or oral briefing by the client on the planned marketing action. This should include subjects such as:

* Planned target group
* Budget or planned production
* Extent and kind of promotion or mailshot (number of pages, with good presentation, coupons, discounts, etc.)
* Involved industries/departments
* Goods involved in the promotion
* Presentation scenario, for example, 'Garden party'
* Transmitted image, for example, aggressive pricing, brand competence or innovation
* Pricing structure

Perhaps the situation is that the planned marketing activity is to reactivate frequent buyers who have not bought during the last year. You have carefully to define who is the planned target group in terms of what is meant by frequent, who is a buyer, do you include those who buy but return the goods, how about those people who have not paid, what goods are included, is there a price/value cut-off, is the channel used important or the location of the purchase, etc. How do we classify buyers who bought frequently 10 years ago but stopped 3 years ago and those who have bought only three times and stopped just recently? These questions are not insurmountable, but they need to be agreed with the client, and the definitions need to be recorded and stored for reference. The main reason is that these definitions affect the target and may affect the model.

The following information is needed:

* Common specification of the aim, for example, turnover activation, reactivation of inactive customers or cross-selling
* Clarification of the different possible applications, for example, to estimate a potential target group or for a concrete mailshot
* Commitment to the action period and application period

- Consideration of any seasonal influences to be noted
- Consideration of any comparable actions in the past

It is of paramount importance to become adept at extracting this sort of information. It is a good idea to practise in a pilot situation to see how best to do it.

Common pitfalls are when the client has not fixed all of the details in time for the initial discussion or when things change between the briefing and the action without the data miner being informed. Sometimes, marketing colleagues prefer not to be seen to be too precise as they may feel that it limits their flexibility. But without knowing all the details, there is little chance of building a good model. For example, the client may say that the action is a branding exercise aimed at bringing people to the usual points of sale, and so the target group is wide and less price oriented including people who are not particularly price sensitive; then, the campaign is changed to being one of aggressive pricing, and the target group is sub-optimal because they are not particularly price sensitive. So the action underperforms.

Experienced data miners find that this problem definition step is decisive in adding value and determining whether they will be successful or not. It can take a long time but is well worth the effort. A bit of psychology can be useful here; see *Caulcutt* references in the Bibliography that look at how data analysts can optimise their interaction with process owners to ensure they really understand each other.

Furthermore, the baseline situation should be evaluated before any data mining work takes place so that the benefits can be assessed. For example, make sure that key performance indicators, such as response rate, cost of mailshots and purchase frequency, are known. Measurable goals should be defined and agreed with management. However, it should be noted that data mining differs fundamentally from hypothesis testing in that it is essentially exploratory so that it is not possible to define exact goals, but we can define the areas where we expect improvement.

2.4 DATA: PROVISION AND PROCESSING OF THE REQUIRED DATA

To provide the required data for the analysis, we must consider the analysis period, basic unit of interest and estimation methods, the variables and the data partition to generate the learning/testing data and for random samples.

2.4.1 Fixing the Analysis Period

In deployment, there is likely to be a time gap between using the model that the data miner has produced and carrying out the activity. For example, we may use our data mining model to determine a mailing list of customers who are most likely to buy, but those customers do not receive the mailshot until a few days (or hours or months) later.

The analysis period consists of the base period (for the input variables) and the aim or target period (for the target or output variables). The base period always lies before the target period and reflects the time gap between running a model and using the results of running the model.

From past activity, we decide how big the deployment time gap is, and then, we include a temporal mismatch in the modelling data so that, for example, input variables such as age, location, segment and purchase behaviour are from not later than one period, say, period number 10, and target variables such as purchasing action or churn activities are from a later period, say, period number 14, and not before. Note that the time period differs depending on the type of business and could represent days, months, quarters or some other time unit of the business cycle.

This temporal mismatch of variables is a major difference from other statistical modelling, but does not present major methodological challenges; rather, it is a question of careful consideration and correct data preparation.

To allow for seasonality, often, a period which lies approximately one year before the application period is chosen as the target period. The corresponding base period is typically a few weeks before the target period and is determined as in the preceding text by practical considerations of time taken to undertake the activity including posting and printing. For example, in the briefing, the client says she wants a Christmas mailing this year, so customers need to receive promotional literature by the end of November. Because of the known seasonality, we decide to use a target period of December 1–31 last year. As it is known that processing and delivery takes four weeks, the end of the base period is 31 October last year. So in preparing the model, we use input variables up to 31 October and target variables for December 1–31. In the application period, the model is used with input variables from the current year to 31 October to determine who should be sent promotional literature this year. We are predicting purchasing behaviour of customers December 1–31 this current year. Note that we have left the whole of November for the processing and delivery of the mailshot.

Besides the temporal shift in the data, the availability of the data also needs to be considered. Typical traps include:

Procedure: Temporal delimitation
Determination of the analysis period consisting of base period and target period

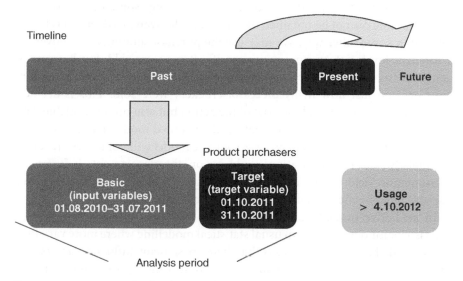

FIGURE 2.4   Time scales for data mining process.

- In the application, one of the necessary datasets is not available yet.
- Departments or industries have changed between the analysis (includes base and target) period and the application period.

We note again that the target variables usually refer to a different time from the explanatory (input) variables. The time scale issue is summarised in Figure 2.4.

2.4.2   Basic Unit of Interest

It has to be decided what constitutes a case, or the basic unit of interest, whether it is an existing person, a company or a location or an invoice, etc. For marketing, the unit is usually a person, because the person is the one making the purchasing decision. But in other industries, a case could be a day (i.e. a case contains a day's worth of data), and base and target periods are probably simultaneous. In a production process, a unit could be all the materials making up a manufactured product, and the target is the quality of the product. The interpretation of the temporal mismatch is that there usually needs to be a run-in period before changes in material input affect the output, and this should be considered in the modelling.

2.4.3   Target Variables

The target variable is fixed according to the briefing. Sometimes, a sensible target variable cannot be obtained directly from the available data and is better derived in some other way. We may use the purchase amount or turnover on a different level, not on how many specific items were sold, but how many generic items were sold, for example, not how many pink cups, but how many cups or even just how many pieces of crockery were sold. This is decided mostly on the briefing and the goal of the action but also on the available data and the amount of existing data that fits the planned model, for example, if a very small number of pink cups were sold, the model will not be very reliable.

Similarly, the summarisation could be on the department, industry or consumption field for the inquiry of the target variable. Note that the target variable must be measurable, precise, robust as well as relevant.

In predictive modelling, less variation in the target variable is preferred. This is a major deviation from usual statistical modelling where more variation is preferred. The reason for this is that there are so many influence factors that cannot be controlled; even if you try your best to match the data, there are always time effects and there is no control data; competitors and the business environment all affect the outcome. Therefore, too much precision is irrelevant and is misleading. In predictive modelling, binary and categorical targets can be quite acceptable, whereas in other statistical modelling, a continuous variable is usually preferable. The statistical reason for avoiding binary or categorical targets is that they require more data; however, with data mining, there is usually a lot of data available, so this is not an issue.

2.4.4   Input Variables/Explanatory Variables

All input variables are generated only for the base period. A subtle but important point is that they need to be used in the data mining process as they were at the end of the base period. This can cause problems with variables that are not static but subject to change, like marital status or address. Unless it is possible to determine if they have changed, these variables should be used with caution, even if they may usually be considered static or slow changing.

More stable, improved models are obtained by classifying continuous variables. When variables such as turnover or purchase amount are classified, it stresses more strongly the differences in the business process. For example, it has quite different implications if someone spends 0 € compared to someone spending 1 €. Mathematically, these quantities are very similar, but in our business application, any purchase, even 1 €, implies some interest in our business, whereas zero really could mean no interest. Without classification, the difference between no purchase and a small purchase would be undervalued. At the other end of the scale, it is more important to know that a buyer belongs to the top 10% of people spending the highest amount than that they spent 2516 € rather than 5035 €. In a sense, the significance of a buyer spending 5035 € is mainly that they are in the high spender category.

The classification can be carried out in a number of ways as described later.

2.5   MODELLING: ANALYSIS OF THE DATA

There are clearly many different data mining methods available, and more are being developed all the time. The core of the data mining process is creating a good model. Good means that it predicts well. However, because data mining is often deployed in a dynamic and variable environment, a fast model for an appropriate business problem generated quickly and deployed accurately and sensitively can have a higher business value even if it predicts slightly less well than a model which takes longer to find. These analysis methods are described in detail in Chapter 6.

Data mining tools are relatively easy to use. It is important to pay attention to the whole data mining process. This includes the steps given previously: problem definition, careful data selection, choice of variables and also checks on relevance and accuracy of models.

There is plenty of data mining software available offering the common (most versatile) methods. Depending on company policy, algorithms can also be written or obtained in freeware. Personal preference may also be for a more visual process-oriented approach requiring minimal programming skills.

Good data mining software should include sound tools for data preparation and transformation. It should be straightforward to obtain deployment models or scripts that can be used easily in a different environment.

2.6   EVALUATION AND VALIDATION DURING
THE ANALYSIS STAGE

The assessment of the quality of the calculated model can be done in three ways: using a test sample having the same split (between target = 0 and target = 1) as the training sample, using a test sample that has a different stratification and using a test sample that has the same split as the whole dataset. We may generate a number of candidate models using regressions, decision trees, etc. The models may differ in terms of the variables included in the model. We then have to compare the models, and this is done by applying each model to the test samples

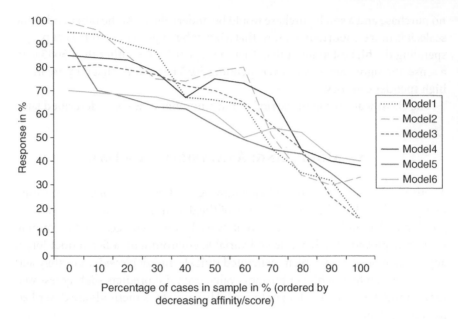

FIGURE 2.5   Lift chart to compare models.

and comparing the results. Some data mining software provides comparisons between models automatically or gives a tool to help compare the models as run on the same samples.

Comparison methods include constructing lift and gain charts and preparing confusion matrices.

Figure 2.5 shows a typical chart.

The best model depends on the business question. Consider the next figure (Figure 2.6) which illustrates two models with quite similar results. If we want good discrimination of the best customers, we choose the dark line model because the first 20% of customers have a higher response. If we are interested in good discrimination for half of the people, then both models are similar. If we are interested in the worst 10%, then again both models are similar.

Sometimes, the lift chart needs to be plotted with a finer scale to see any abnormal areas. For example, there may be an unstable area around 40%, and if that is an area of interest, then the models should not be used.

Figure 2.7 shows model 1 from Figure 2.5 in finer detail. It has three unstable areas in the middle. However, if we just need the top 20% (or bottom 40%) of cases, the model is still stable enough to use.

An additional model comparison is given by the confusion matrix. A good model has similar sensitivity in the training and testing phases. In Figure 2.8, the values are similar which is good. A slight difference can be OK, but a model with a big difference is not desirable.

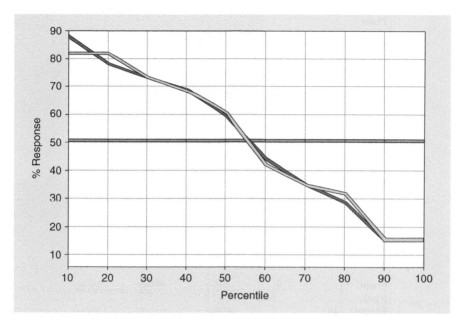

FIGURE 2.6   Lift chart to compare models.

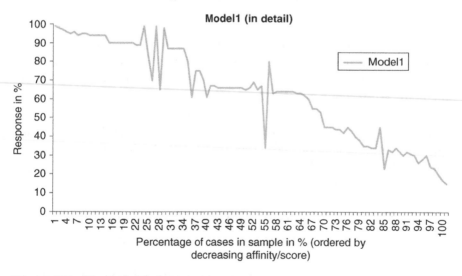

FIGURE 2.7   Fine scale lift chart.

An alternative tool using Excel is shown in Figure 2.9.

Sometimes, the ability of the model to rank the customers in a relevant way is more important than the statistical quality of the models. A useful model is one that gives a credible rank ordering of customers in terms of associated variables.

Train		To (predicted target)		
		1	0	all
	1	**5040**	960	6 000
From (real Target)	0	1080	**4920**	6 000
	all	6120	5880	12 000

Validation		To (predicted target)		
		1	0	all
	1	**4860**	1140	6 000
From (real Target)	0	1020	**4980**	6 000
	all	5880	6120	12 000

FIGURE 2.8 Confusion matrix for comparing models.

Another way to validate the models offered by most data mining software is cross-validation. This is a way of assessing how the results of the analysis will generalise to an independent dataset and includes a number of different methods. We recommend that cross-validation is used when the dataset is small or contains highly reliable, high-quality data.

The most important aspect of validation is to check that the model or any other data mining solution makes sense as far as the business is concerned and that the results are credible and usable for the benefit of the business.

2.7 APPLICATION OF DATA MINING RESULTS AND LEARNING FROM THE EXPERIENCE

The point of doing data mining is to use the results and move into action, for example, to:

- Find the best customers, for example, as a blueprint for brochure distribution or mailshot
- Score relevant factors of influence which describe the target group

Scoring results: determination of edition

Assumption: Costs per unit 0,70 €, adjustment factor response 0,3, estimated revenue per customer 12 €

Rating after scoring		Real results of the analysis inventory						Expected results of future action		
Group	Number of Customers	Nume of Customers cumulated	Number of buyers in departments	Buying Rate	Cumulated Byirg Rate	Revenue (analysis period in departments)	Estimated cost per group	Expected buyers adjusted by factor	Expected buying rate adjusted by factor	Expected revenue per group (real av. Rev.. Buyer * exp. Buyer)
1	50.000	50.000	20.604	41,20%	8,25%	328.322,0	35.000	6.181	12,36%	74.174,40
2	50.000	100.000	16.721	33,44%	14,95%	246.750,0	35.000	5.016	10,03%	60.195,60
3	50.000	150.000	10.766	21,53%	19,26%	152.848,0	35.000	3.230	6,46%	38.757,60
4	50.000	200.000	9.107	18,21%	22,91%	132.450,0	35.000	2.732	5,46%	32.785,20
5	50.000	250.000	9.025	18,50%	26,52%	125.896,0	35.000	2.708	5,42%	32.490,00
6	50.000	300.000	8.933	17,87%	30,10%	127.080,0	35.000	2.680	5,36%	32.158,80
7	50.000	350.000	8.007	16,01%	33,31%	112.690,0	35.000	2.402	4,80%	28.825,20
8	50.000	400.000	8.134	16,27%	36,57%	117.566,0	35.000	2.440	4,88%	29.282,40
9	50.000	450.000	8.156	16,31%	39,83%	115.885,0	35.000	2.447	4,89%	29.361,60
10	50.000	500.000	7.209	14,42%	42,72%	100.967,0	35.000	2.163	4,33%	25.952,40
11	50.000	550.000	6.624	13,25%	45,38%	94.278,0	35.000	1.987	3,97%	23.846,40
12	50.000	600.000	5.516	11,03%	47,58%	67.026,0	35.000	1.655	3,31%	19.857,60
13	50.000	650.000	5.496	10,99%	49,79%	66.916,0	35.000	1.649	3,30%	19.785,60
14	50.000	700.000	5.657	11,31%	52,05%	69.432,0	35.000	1.697	3,39%	20.365,20
15	50.000	750.000	5.543	11,09%	54,27%	67.842,0	35.000	1.663	3,33%	19.954,80
16	50.000	800.000	5.585	11,17%	56,51%	69.919,0	35.000	1.676	3,35%	20.106,00
17	50.000	850.000	5.262	10,52%	58,62%	63.706,0	35.000	1.579	3,16%	18.943,20
18	50.000	900.000	4.380	8,76%	60,37%	60.561,0	35.000	1.314	2,63%	15.768,00
19	50.000	950.000	4.429	8,86%	62,14%	60.447,0	35.000	1.329	2,66%	15.944,40
20	4.616.931	5.566.931,00	94.511	2,05%	100%	1.137.509,0	3.231.852	28.353	0,61%	340.239,60
Gesamt	5.566.931		249.665	4,43%	100%	3.318.060	3.896.852	74.900	1,35%	898.794,00

Very good

Recommended edition

Very bad

FIGURE 2.9   An example of model control in Excel.

While it is acceptable when finding the best customers to look only at the training analysis period, the application of modelling results to a planned future period must be carried out thoughtfully.

All variables must be transferred across the time period between analysis and time of application, and customer scores must be determined on the basis of the variables current at that time. For example, consider the variables age and lifetime revenue. If the period between analysis and application is one year, then at the time of application, we would transform the age variable by creating a new variable that represents the corresponding age of the person during the training analysis period, in this case by subtracting one year from the age variable or by recalculating the age from the birthdate if that is available. The score for that person is then calculated using the new age, that is, after subtracting one year. For the case of lifetime revenue, we cannot use the current value at the time of applying the model but have to recalculate it by summarising the revenue up to the training analysis period.

A point to note is that information available in the training analysis period may not be available at the time of applying the model. If this problem is known in advance, then variables related to that information should be omitted from the model. If a variable concerns a specific feature that we know is likely to change, then it should be replaced with a more generic variable. For example, consider the purchase of a yellow pen. If we know that at the time of application, yellow pens will not be sold as they will no longer be fashionable, then we would be well advised to replace 'yellow pen' with a more generic variable, such as 'any pen'. Another possibility is to replace the variable with 'purchase of a fashionable pen' without specifying the colour as this information is likely to be available at the time of application. The new variable must be created in the dataset so that a value is available both in the training and in the application period.

The success of the predictive modelling must be assessed. This is usually done by comparing predictions with outcomes. However, we need to look at the whole process to ensure that it has been worthwhile. Data mining is a costly process for the company, and so managers will expect substantial benefits to arise from the expense. The improvement on baseline key performance indicators needs to be clearly communicated. Lessons learnt from the process should be fed back to ensure a continuous cycle of improvement.

PART II

DATA MINING PRACTICALITIES

All about Data

A Practical Guide to Data Mining for Business and Industry, First Edition.
Andrea Ahlemeyer-Stubbe and Shirley Coleman.
© 2014 John Wiley & Sons, Ltd. Published 2014 by John Wiley & Sons, Ltd.
Companion website: www.wiley.com/go/data_mining

3.1   Some Basics

In most companies, marketing, sales and process control are major drivers for promoting data quality and producing comparable numbers and facts about the business. But, even production and Research and Development (R&D) departments need reliable data sources to use statistical methods or data mining to improve output and profitability. In most companies, the main impetus for checking and restructuring the data and processes is the introduction of Customer Relationship Management (CRM). CRM imbues the company's own data with new meaning. Gone are the days when customer data management only meant using the correct address in the mailing list. Today's data management must target individual customers and provide more communication and better quality information tailored to these specific customers and their customer behaviour. Data management forms the basis for using intelligent methods such as data mining to analyse the wealth of knowledge available in a company and build an optimal communication with customers and stakeholders.

Concept	Interpretation	Relationship
Wisdom	Applied knowledge	
Knowledge	Information in context	Conceptual umbrella for information and data
Information	Meaningful data, data in context	Knowledge needed for special purposes
Data	Representation of facts	Fundamental

FIGURE 3.1 Important terms in data evolution.

In many companies, there is hardly any distinction between the terms knowledge, information and data. Among other things in computer science and economics, it can be seen in the literature that there are strongly divergent views and that different approaches also exist within the two specialties. In computer science, the terms information and data are often used interchangeably, since an explicit distinction does not seem absolutely necessary. The data is equated with the information it represents. The economist, however, sees information as a significant factor of production as well as the intermediate or final product of the corporate transformation process. Information and data are distinct for the economist. This divergence of views between computer science and economics has implications for how different specialties view data preparation. Computer scientists sometimes miss information that is not coded in the data directly, whereas economists are more familiar with using additional knowledge not stored in data systems (see Figure 3.1).

3.1.1 Data, Information, Knowledge and Wisdom

Here, we demonstrate a clear distinction between knowledge and data. The starting point is often the definition of information as knowledge needed for special purposes.

An early indication of the transition implied by Data, Information, Knowledge, and Wisdom (DIKW) is given in the T.S. Eliot poem *The Rock* (1934) where this hierarchy was first mentioned:

Where is the life we have lost in living?
Where is the wisdom we have lost in knowledge?
Where is the knowledge we have lost in information?

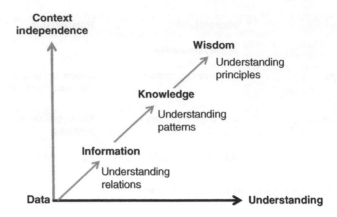

Figure 3.2 The evolution of wisdom. Source: Reproduced by permission of Gene Bellinger (http://systems-thinking.org/).

DIKW is illustrated in Figure 3.2.

D = Data which is facts without context
I = Information which is facts with some context and perspective
K = Knowledge which is information used to detect and understand patterns
   in the data
W = Wisdom which is knowledge and happens when you understand why the
   patterns are occurring

It may seem unusual to write about knowledge management and data theory in an applied book for data mining, but this viewpoint may help to understand how knowledge that may or may not be part of the data itself can and should be included in data preparation. For example, it may be known that a company has a seasonal competitor that affects their sales periodically, or blips in production may be caused by a known fault.

3.1.2   Sources and Quality of Data

Data to be used for enterprise information and knowledge creation can come from either internal or external sources (see Figure 3.3). The operational information system moves the large amount of data produced internally through the various processes. Since internal data is used primarily to handle the daily business, the operational systems lack any facility for keeping a comprehensive history. Inconsistencies may arise because of partially duplicated data storage in the very different sub-systems. Just as many quality

Data source	Example	Characteristics
Internal	Date a product was manufactured or invoice data	In control of company and may be more reliable
External	Credit rating or data about the area the customer lives in	May not be a perfect match in time scale or location

FIGURE 3.3   Typical internal and external data in information systems.

defects affect the data used in the operational systems, so quality defects have an even greater impact on the analysis-oriented information systems. The quality of the data has a significant influence on the quality of the analysis based on it. At least the quality and reliability of internal data is in the control of the company. This is not the case for external data.

External data is generated outside the company's own processes; it is often required to act as additional information (e.g. credit rating) or as reference values (e.g. data from Government Statistical Offices or National Statistics Institutes). For analytically focused information systems in areas such as Database Marketing (DBM) and Customer Relationship Management (CRM), external data is added frequently; there may be specifically purchased additional information about the customer or the customer's address.

Often, the quality of internal data is better than that from external resources, not least because you can control exactly how and when the internal data was generated. Another issue with external data is that it may not match the internal data exactly in time (being contemporaneous) or location. These discrepancies should be noted, but usually, even poorly matched external data can be useful to supply additional relevant information.

3.1.3   Measurement Level and Types of Data

There are different types of quantitative data, all of which can have good information content. There are many different terms used to describe the different data types, and the most common words are detailed in the following.

The simplest level of measurement is expressed as nominal data which indicates which named category is applicable. For example, a customer may live in an urban area, a rural area or a mixed area. This nominal data variable will consist of a column of urban/rural/mixed with one row for each customer. If there are only two levels, for example, 'buy' or 'no buy', then the data is referred to as binary variables. If there is any order associated with the categories, then they are referred to as ordinal data. For example, text associated with reasons for returning goods may read something like:

The clothes were the wrong size.

This comment could be classified as a size complaint. The frequency of size complaints could be compared with the frequency of non-size-related complaints. Size/non-size related is a binary variable having two levels, and we could usefully compare the number of complaints at each level.

If the reason for returning is

The clothes were too big,

then we could classify this complaint as a 'too big' mismatched size complaint, and we could compare the frequency of 'too big' with 'too small' mismatched sizes or 'non-size' related. A variable containing the information about complaints classified as too big/too small/unspecified size/non-size related is a categorical variable with nominal measurement at four levels.

There could also be an ordinal-level measurement if the categories are related in a descending or ascending order. For example, the variable levels could be first return, second return, third return, etc. If there are more than two levels for a nominal variable but there is no implied order, then some data mining procedures may require them to be converted to a series of indicator variables. For example, urban/rural/mixed could be converted to three indicator variables: urban or not, rural or not and mixed or not. The last variable is redundant as its value is indicated when neither of urban nor rural is true.

Variables that represent size are referred to as measures, measurements or metrics and are described as being on a metric level. In data mining, the term 'metric' includes counts of some data type, like page views, and may correspond to a column of data.

The measurement level would be interval if the variable is the number of occurrences, for example, the number of returns for a customer (i.e. the number of times a customer returned an order). In this case, there could be lots of customers with zero returns but a few customers with one, two, three or more returns. This is discrete data measured on an interval scale. Another example of interval-level measures or metrics is provided by *altmetrics*, which are measurements of interaction based on the social web resulting in variables like the number of hits or mentions across the web. Subjects such as netnography explore web activity in great detail.

Many data items are measured on a continuous scale, for example, the distance travelled to make a purchase. Continuous data does not need to be whole numbers like 4 km but can be fractions of whole numbers like 5.68 km. Continuous data may be interval type or ratio type. Interval data has equal intervals between units (e.g. 3.5 is 1 less than 4.5, and 4.5 is 1 less than 5.5). Ratio data is interval-type data with the additional feature that zero is meaningful and ratios are constant (e.g. 12 is twice as big as 6, and 6 is twice as big as 3).

Nominal and ordinal variables are referred to as categorical or classification variables. They often represent dimensions, factors or custom variables that allow you to break down a metric by a particular value, like screen views by screen name.

To summarise, in data mining, we consider classification or categorical variables which can be nominal, binary and ordinal as well as scale or metric variables which can be count, continuous, interval or ratio.

Qualitative data, such as pictures or text, can be summarised into quantitative data. For example, an analysis of content can be expressed in terms of counts and measured in terms of impact or quantity of relationships. Content analysis may give rise to nominal data in which the categories can be named but do not have any implied order.

See Bibliography for texts including discussions of data types, qualitative and quantitative data and information quality.

3.1.4 Measures of Magnitude and Dispersion

The measures of magnitude used in data mining are mainly the arithmetic mean, the median and the quantiles. Consider a simple situation where the expenditure in € of 10 people is as follows: 34, 21, 0, 56, 24, 26, 12, 0, 14, 33 (see Figure 3.4). The arithmetic mean is the sum of the values divided by the number of values, so it is 220/10 = 22.0. The arithmetic mean is usually what is meant by the average. In the aforementioned example, each expenditure value is equally

Custo-mer	1	2	3	4	5	6	7	8	9	10	Total
Expen-diture	34	21	0	56	24	26	12	0	14	33	220
Weight	10	20	5	15	20	5	5	5	5	10	100
Product	340	420	0	840	480	130	60	0	70	330	2670

FIGURE 3.4 Table of sample data.

important, and so the average value is just the sum divided by 10. However, sometimes, the importance of data items may vary, and the weighted average value may be considered. For example, each of the expenditure values represents different segments of customers, and the segments have the following % weights: 10, 20, 5, 15, 20, 5, 5, 5, 5, 10. In this case, the weighted average is the sum of the weights times the expenditure divided by the total weight, which is 2670/100 = 26.7. The calculation is shown in the table.

In our example the weighted average is higher than the average because the larger expenditure values generally occur in the more populated segments.

The median is the central value when the data is placed in numerical order, so in our example, it is midway between the fifth and sixth values. Placing the 10 values in order gives the following: 0, 0, 12, 14, 21, 24, 26, 33, 34, 56. The median is (21 + 24)/2 = 22.5. The median has half of the values below it and half above it. There are four quartiles, and the first quartile has 25% of the data below it; the second quartile is the median and has 50% of the data below it; the third quartile has 75% of the data below it. The first and third quartiles are sometimes referred to as the lower and upper quartiles, respectively.

Quartiles are a particular type of quantile. We can define other quantiles and will make use of sextiles which divide the data into six parts. Quantiles are usually calculated from sets of data with many more than 10 members.

The measures of magnitude can be used to assess the distribution of interval (metric) data; if the difference between the median and the average value is large, then this indicates that the data variables:

- Are not Normally distributed
- May be skewed
- May have one or more outliers

If there are outliers, then they must be considered in the analyses accordingly or be excluded. If Normal data is required, then a transformation may help.

The dispersion measures show how variable the data values are. The dispersion measures commonly used in data mining are the range, inter-quartile range, variance and standard deviation. The range is the difference between the largest and smallest data values. In the aforementioned example, the range is 56 as this is the difference between 0 and 56. The inter-quartile range is the difference between the first and third quartiles. The variance is the square of the standard deviation, and both are important measures as they relate to calculations used in the Normal distribution. See standard statistical textbooks in the Bibliography for the relevant formulae.

3.1.5   Data Distributions

Data mining is carried out on data collected for many people or cases. The way a data item varies is referred to as its distribution. The variation of categorical data can be shown by the frequency of occurrence of each level either in a table or in a bar chart which shows how the responses are distributed across the collection of cases (see Figure 3.5).

Histograms are used to show the way scale data is distributed. Data, like salaries or customer lifetimes, are asymmetric with most values being below the average and a few values being much higher. Typically, the average salary will be much higher than the median salary because the few very rich people give the salary distribution a positive skew.

A commonly occurring histogram shape is where most observations are close to the average and fewer cases are further away either larger or smaller than the average. This is the shape of a histogram for Normally distributed data; it is symmetrical and bell shaped. Measured data, such as a person's weight, usually has a Normal distribution. The Normal distribution also arises when average values are plotted instead of individual values. For example, if average customer lifetimes are calculated for random samples of customers, then a histogram of averages will probably have a Normal shape. The larger the number in the samples, the closer the average values will be to an approximately Normal distribution.

3.2   Data Partition: Random Samples for Training, Testing and Validation

There are usually more than enough customers (or cases) available for almost all questions in the analysis. We should use representative random samples for the analysis not only because it speeds up the calculations for the modelling but also because we can then test and validate the models and be more confident that they are robust to changes in the population which may be caused by changes in the process, the business, the environment or any other focus or time effects.

We recommend that several samples are generated for training, testing and validation. If the Database (DB) is big enough, then the samples can be selected without replacement; in other words, each case can only be picked once. However, if the DB is small and especially if the target group of interest is small, then we can sample with replacement, which means that cases have the possibility of appearing more than once.

In summary, the model is generated (or learnt) using the training (learning) sample. We then proceed to apply the model to the testing sample. We may then try the model out on further validation samples.

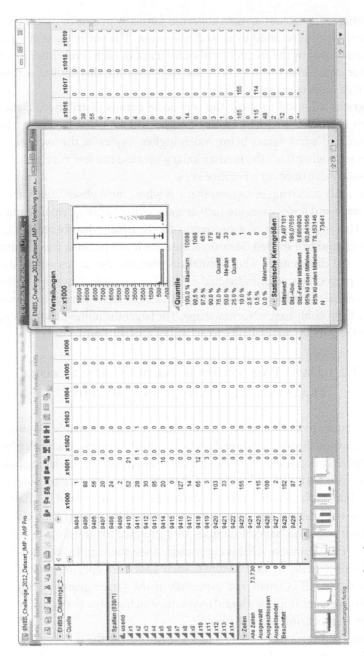

FIGURE 3.5 Data distribution.

If the DB is large enough (100 000 or more), then in practical terms, we have found that a random sample size of approximately 30 000 works well. If the target group of interest is only represented by a few customers, then a simple random sample will contain very few of the target cases who are, in fact, the ones whose behaviour we want to understand. Therefore, we will probably need to carry out a stratified random sample in which random samples are taken from each of the strata given by the target variable. In practical terms, this raises these issues:

- What should the split be between strata in the sample?
- How can we get a large sample from a small population?
- If the split is not the same in the training sample as in the population, will the model fit the real population?

Deciding on the split depends on the relative and absolute numbers of target and non-target cases. As a rule of thumb, our experience suggests that if the proportion of target cases is less than 1% and the number of target cases is less than 15 000, then a 1:2 or 1:3 split should be used. This is because choosing only 15 000 from the much larger population of non-target cases risks missing some of the more unusual cases, whereas three times (45 000) lessens this risk (see Figure 3.6).

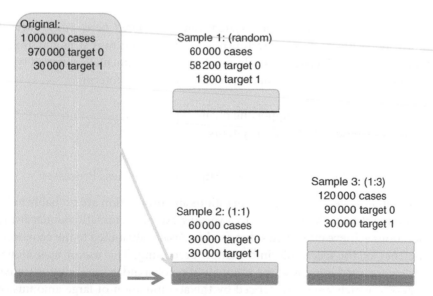

FIGURE 3.6 Stratified sampling.

If the number of target cases is small, then we may choose to augment it by sampling with replacement. This is a form of bootstrapping and raises the complication that the same cases can appear multiple times in both the training and testing samples. The theoretical implications of this bootstrapping are uncertain; however, in practice, building a model using the larger samples has been found to produce a stable model and is preferable to using the smaller target population as the sample in the traditional way.

For further testing and validation, we can use other stratifications than 1:2, 1:3 or 50/50 or a simple random sample which represents the original proportions of the binary target variable in the population. It is important that the model gives good, reliable results with the actual data, so a simple random sample which reflects the actual data gives a good test of the model. Note that the original model is generated on the stratified sample to make the modelling process more stable. However, this is an artificial world, and we need to check that the model still applies in the more real situation.

Note that some data mining software offers the option of data partitioning in which the user states proportions, such as 60% and 40%, for the training and testing samples. In this case, we can present the full dataset to the model, or we can prepare our stratified sample and present that to the software.

3.3   Types of Business Information Systems

Only computerised information systems are considered here; other internal information systems, such as meeting notes, notices, hearsay, post-its and similar items, are not considered further. In practice, two fundamentally different types of information systems have become established in the enterprise:

- Operational systems supporting business processes
- Analysis-based information systems

3.3.1   Operational Systems Supporting Business Processes

Information systems to support operations and production are probably to be found in virtually all businesses to a varying extent. Automatic standardising operations achieve efficiencies that can be causally attributed to the economic efficiency of the electronic information processing. The use of operational systems is aimed initially at the rationalisation of standardised administrative processes, which are characterised by the accumulation of large amounts of data, resulting in the shortening of throughput times of processes. Operational information systems consist of the sum of all the individual systems which are necessary to establish the daily business. Part of the operational information systems in the traditional Enterprise Resource Planning (ERP) systems, including financial accounting, are the data acquisition systems. All these systems have in common the fact that they have been optimised to specific aspects of operational work, both in data processing and in the amount of data being stored. The DB may be implemented, but its maintenance is not aligned with all the uses of the data (leading to redundancy or duplication of effort). For example, amendments to addresses may be made for the benefit of one process but not be transferred to apply elsewhere. That and the fact that such systems are often composed of very heterogeneous DBs implies a risk that inconsistencies may arise. Another feature of operational systems is the method of data processing the large numbers of individual records. These methods are different to those employed in the second type of information system, the analysis-based information system considered in the following.

3.3.2   Analysis-Based Information Systems

Analysis-based information systems refer to all systems that are necessary to store the data and make it ready for analysis. They also include the (user) tools, with whose help one can gain benefit from the information and knowledge. In terms of schematic architecture, the connected front-end tools that help the user to access the datasets can be distinguished from the data storage components. The front-end tools are thus not part of the data warehouse in the strict sense. Ideally, there should be independence between the storage and the various tools (including data mining tools) that have access to it and the interfaces that can exchange data.

Despite this conceptual separation, it should not be overlooked that in practice there is often a close integration of technological tools with the data storage components. Specifically, analysis-based information systems consist of data warehouses and data marts as well as analytical tools (i.e. Online Analytical Processing (OLAP) and data mining tools). Data warehouses and data marts are discussed in detail in the following sections. Further details of analytical tools and a checklist for choosing and using them are given in later chapters and as they arise in the context of the data mining recipes in Part III.

3.3.3   Importance of Information

Information plays a crucial role in a company's competitiveness. Companies that use innovative technologies are able quickly and flexibly to adapt to rapidly changing market factors and customer requirements, enabling them to achieve a strong competitive advantage. There is a flood of data in the operational DB of the daily business. However, this data capital is often poorly utilised or is left lying idle.

Data capital:

- Is the value contained in the data
- Can only be assessed when data is turned into information
- Is often poorly utilised or is left lying idle

Essential information is often not available to decision makers at management level at the critical point of need or at least is not in a form necessary for creative analysis and decision making. It can be shown that for the DB of these systems, the well-known 80/20 rule applies, in other words that 20% of the data gives 80% of the information needed for decision makers, and as a complement, 80% of the data collected is useful for only a small (e.g. 20%) amount of decision making. The company is typically storing a lot of data (e.g. in tax returns) that will not fully be used for decision making. As a sound basis for corporate decisions, it is vital that meaningful data can be selected and made available quickly. For example, it should be no problem to be able to find the number of new customers or prospects or their mean age whenever it is needed. Having data available to answer the questions relevant to decision makers represents a major strategic advantage.

Examples of questions relevant to decision makers are:

- Which customers should be made a particular offer?
- Which customers are at risk of leaving us (i.e. churning)?
- How high is the cross-selling potential for a new product?
- What is the lifetime profit that can be achieved with which customers?
- How can top prospects with high lifetime values be attracted?
- What is the turnover that can be achieved in the next year?

The fundamental point is: Why is it so difficult to find answers to these critical management questions and why are these questions answered so infrequently? One explanation is that the answers are not straightforward. Considering the nature of the questions listed, the answers do not lie in a single set of figures or customer features, but in the right combination of a variety of different bits of information. Thus, for example, the affinity of a customer to an offer depends on characteristics such as age, gender, marital status, demographic typologies, previously purchased products, interest shown in the product, payment practices and many other properties.

3.4 DATA WAREHOUSES

A data warehouse is a collection of data which is gathered for the specific purpose of analysis. The data warehouse is very different to all other information systems in a company, as the relevant data is quality checked and then possibly processed within the data warehouse. Information systems typically do not allow an overview, whereas data warehouses are designed with this in mind.

Unlike in other operational systems, the mapping of historical data, data history and external data constitutes a large role in the data warehouse. The term data warehouse is now generally understood to mean something that serves as an enterprise-wide DB for a whole range of applications to support analytical tasks for specialists and executives. The data warehouse is operated separately from the operational information systems and filled from internal DBs as well as from external sources of data. The data warehouse is a logical centralised resource.

The term data warehouse is generally understood to imply topic-oriented data rather than a concrete DB system product, with separate company-specific applications; it embraces the underlying concept of combining decision-related data. In other words, whereas other DBs are specific to particular software, the concept of data warehouse depends completely on the ideas that the company wants to explore. It cannot be built up mechanically by software alone.

The contents of a data warehouse can be characterised by four main features, which reveal the significant differences to other operational data:

- Topic orientation
- Logical integration and homogenisation
- Presence of a reference period
- Low volatility

These topics are dealt with in the following sections.

3.4.1 Topic Orientation

In contrast to operational systems, which are oriented towards specific organisation units, remits and work processes, the contents of the data warehouse are oriented towards matters which affect the decisions made by the company in specific topic areas. Typical topics include the customer, the products, the payments and the advertising or sales campaigns. Besides affecting the content of the data warehouse, this fact also has great influence on the logical data model applied. For example, in operational systems, the customer as such does not appear; rather, it is the product and invoice numbers which mainly feature and are followed up in subsequent processes. The data may often be stored in totally different places, including accounting systems, logistics and delivery and stock control. By contrast, the data warehouse will be customer oriented, and if you follow the customer number, it is easy to find all the information associated with this customer, regardless of which system the data is stored in.

3.4.2   Logical Integration and Homogenisation

A data warehouse traditionally consists of common data structures based on the ideas of relational DBs, but nowadays, the discussion of unstructured no-SQL DBs is considered alongside the discussion of 'big data'. The really big amount of data coming from log files and social networks necessitates a different architecture and way of storing data. The aim of both data warehouses and big data architecture is an enterprise-wide integration of all relevant data into a consistent set of data in a continuous system model. This goal also implies the cross-functional use of the data.

3.4.3   Reference Period

Information for decision support should be provided quickly and in a timely fashion. However, it is relatively unimportant for data processing to be taking place at a precise time in the operational systems. It is much more important to be able to incorporate the different periods of time without problems in the ensuing analysis. In the operational system, the time factor has only a descriptive role, but in the data warehouse, it is an important structural component. The special feature of the data warehouse is the fact that historical data is retained, even, for example, data that was archived in the operational system for a long time or was destroyed in a company reorganisation.

3.4.4   Low Volatility

Data that was once stored in a data warehouse should not change although there may be amendments to the data warehouse as a whole. For example, where there are errors in the data due to a faulty charging process, the action is to insert new records to describe the new charging process rather than overwrite the old records. This is in clear contrast to operational systems. As an example, a product might be purchased, but later, the purchase is cancelled. In the operational system, the record with the order would be overwritten by the reversal record or be deleted. In the data warehouse, there would be two records: one with the order and one with the cancellation. Both records would be included in the data warehouse to understand the action of the customer and also to allow different analyses to be carried out. In the operational system, there would be no extra external data or information available to explain the entry or deleted records, since the failed purchase is no longer relevant.

3.4.5 Using the Data Warehouse

In summary, we can say that the data warehouse is a central storage DB, with the aforementioned characteristics, centralising all relevant data in the enterprise. Generally, therefore, the data warehouse concept is characterised by the specific architecture or implementation of form that generated it. This idea of form integrates operational data from which the data warehouse is filled regularly, but thereafter, acts independently from the operational system. The main roles of the data warehouse are to support decision making and for analytical purposes.

Unlike operational systems, a data warehouse is able efficiently to have 'read access' to large amounts of data designed in complex structures. Particular attention is paid here to the changing information needs. Through this, it is necessary to design the structures so that complex queries that involve large amounts of data, as well as extensive aggregation and joining operations, can be managed. This typical form of a data warehouse leads to utilisation which differs significantly from that of the operational systems. In the data warehouse, utilisation is subject to significant fluctuations with pronounced peaks, which are directly related to the queries being made. In contrast, the utilisation of an operational system is close to constant and stays at a uniformly high level (see Figure 3.7).

The construction of a data warehouse helps some companies solve the massive resource conflict between the execution of daily business and the implementation of complex analysis needed to support decision making. To implement a data warehouse, we need to consider three different forms of organisation:

- Central DB warehouse
- Distributed DB warehouse
- Virtual DB warehouse

The most common form of implementation is the central DB warehouse; this is where the management of all datasets for the various front-end applications is on a single system. Distributed data warehouses are when different departments might run their own data warehouses optimised for their needs.

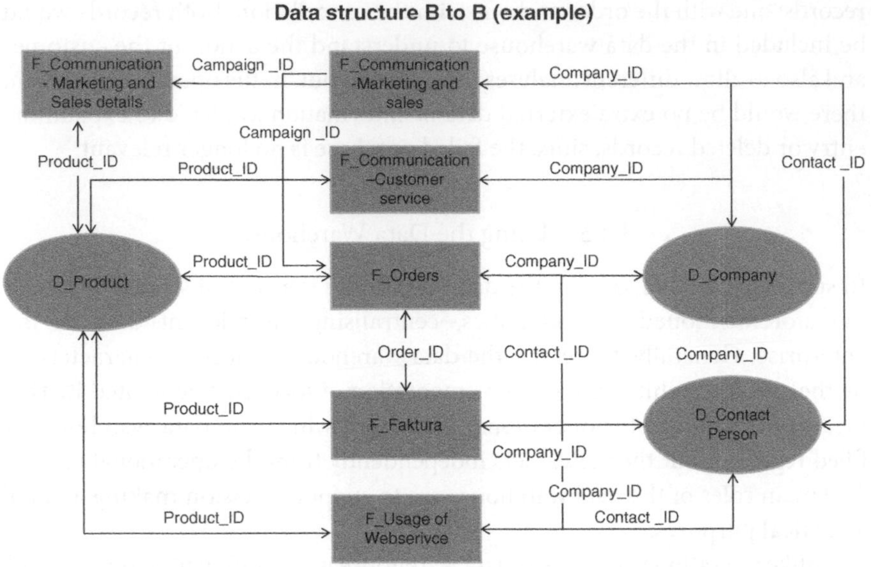

Data structure B to B (example)

FIGURE 3.7 Example data structure.

In some parts of the literature, these are also called data marts. There is confusion in the literature as some people call them data warehouses; there is some inconsistency of the terms used. When we speak about data warehouses, we mean storage of detailed data, and when we speak about the data mart, we mean more or less prepared data for special usages (e.g. data aggregated to control marketing campaigns or to prepare data for data mining).

A virtual data warehouse is sometimes created for reporting and refers to creating views of the original data in the legacy system or the original data sources. Everything done in the central data warehouse by the Extraction, Transforming and Loading (ETL) processes is implemented in the view but not carried out on the real data.

3.5   THREE COMPONENTS OF A DATA WAREHOUSE: DBMS, DB AND DBCS

A data warehouse can be seen as a DB system sharing the three components of the Database Management System (DBMS), the DB and the Database Communication System (DBCS). For example, the DBMS contains meta-data on loading, error detection, constraints and validation, and the DB is the storage of the data; DBCS refers to the possibility of analysing the data using, for example, SQL or other suitable languages.

3.5.1   Database Management System (DBMS)

The DBMS in the data warehouse is mainly for managing the analysis-oriented DB. It provides the functionality for data definition and manipulation; thus, the DBMS in a data warehouse has different requirements to that of an operational system.

3.5.2   Database (DB)

The issues of integrity and consistency in the DB datasets under analysis are evaluated differently in the operational system, as are data security and availability. This is because the data in a data warehouse is made up of copies of operational datasets with the addition of any changes made in the operational system and any additional information extracted from the data. For example, the operational data may store salutation only (Mr, Mrs, Ms, Miss, Master); in the data warehouse in addition to the salutation, marital status could also be extracted as well as gender out of the salutation. If the salutation changes from Mrs to Ms, then the operational data just records the new value, but the warehouse records the old and the new as well as the time it changed, as well as any consequent information like marital status.

Hence, the effort involved in ensuring integrity, consistency, security and availability is greater in the data warehouse than in the operational system. For this reason, only the administrator or defined processes can change the data in a data warehouse, whereas any of the human operators can add to the data in an operational system or change it (e.g. by adding new addresses).

Data is more business oriented in a data warehouse. This can lead to critical and strategic information showing up earlier. There is a greater demand to think about the issue of security and the consequent roles of data because the data is now immediately useful, say, to competitors.

Storage and access must be optimised to give the user a short response time for complex queries and analysis, but this must not lead to a loss of flexibility in the analysis.

3.5.3   Database Communication Systems (DBCS)

DBCS play a prominent role in analysis-oriented information systems, because without them, the use of the data stored in the data warehouse is very difficult. At the same time, front-end tools and their internal data management systems put very different demands on the interfaces.

An indispensable part of the analysis-oriented information system and particularly of the data warehouse is detailed meta-databases. Unlike in the operational system, where their role is less important, meta-databases are particularly suitable for the users of data warehouses because the meta-data is essential to perform the analysis on the data effectively.

Evidently one of the critical success factors for an analytically oriented information system is a well-maintained meta-database complete with the relevant business terms.

3.6 DATA MARTS

The term data mart is widely used and is well differentiated from the term data warehouse. However, both data warehouses and data marts are building blocks that serve to store data in the context of analysis-oriented information systems.

A data mart is defined as a specific collection of data, in which only the needs of a specific view and use are mapped. For example, a data mart could be constructed for customer-based predictive analyses such as the prediction of those customers with the highest probability to buy next.

Data is often arranged in very different ways. A data mart on the one hand is seen as a subset of the data warehouse, in which a portion of the dataset is duplicated, and on the other hand as an entity in its own right. If there is duplication of data in the data mart, this is justified by the size and structure of the data warehouse. The data warehouse contains very large datasets that are based on 'relational' DB systems and are thus organised in relation to usage; they are not necessarily structures which are fully adequate for addressing specific problems. Especially when interactive access to the datasets is desirable, the representation of data in the data warehouse as well as the response times may not be very good. Constructing a data mart solves this problem; function- or area-specific extracts from the data warehouse DB are collected and stored in duplicate in a data mart.

Data mart storage can be realised with the same technology, and a data model can be used that corresponds to a proper subset of the data warehouse, so that the data mart can be easily maintained. Alternatively, it also seems appropriate for the data mart with its manageable data volume (as opposed to the relational-based data warehouse) to use a multi-dimensional DB system in order to exploit the potentially better modelling and querying capabilities of this technology. In particular, the necessary transformation of data into the new model can be carried out. However, because the care of such data marts is expensive, it is sensible to consider the advantages and disadvantages of heterogeneous data models.

Note that the term 'models' here is used in the sense of computer science and refers to the data structure including aspects such as the following: which data item is the primary key and which is the secondary key?

The users receive the data mart tailored to their information needs and including a sub-section of the enterprise-wide DB. With careful delineation of these data mart excerpts, requests for essential parts of the data mart can be compared favourably in terms of speed of access as compared to the speed of direct access to the data warehouse. Basically, data marts can be close to the form of data storage (relational and multi-dimensional) in the data warehouse, but they differ in that unlike the data warehouse which is created only once, the data mart is regularly updated as a whole or in part.

3.6.1   Regularly Filled Data Marts

Data marts, which need to be updated regularly according to their initial load, are often needed for reporting and OLAP or when you are data mining aggregated data that is continuously available. Typical examples of such data mart tables and files are compressed at different levels (e.g. sales figures for the current year). Depending on the definition of this information, it should be available daily, weekly or monthly. The shorter the update period, the more important it is that the process of updating is fully automated or that the refill is carried out according to fixed rules and within defined procedures. With a daily loading cycle, updates take place during the night, after data from operational systems have accumulated in the data warehouse, giving the most current information available.

3.6.2   Comparison between Data Marts and Data Warehouses

Many companies offer help with data marts to provide users with information pre-aggregated. The data warehouse DB is stored there with the current and historical data from all divisions in the different stages of compression in the core of the analysis-oriented information system. Here, there is a recognisable conflict from the user perspective: in a data mart, data is aggregated for the analysis of major interest; however, there is little flexibility to analyse detailed individual values or to respond to new data requirements to link them.

Data warehouses do not give rise to this dilemma; data marts may be deployed to contain aggregated and possibly transformed data, while in the data warehouse, the data is stored on the finest available granularity. For the data warehouse, the relational storage of data as a quasi-state has emerged over the years, while data marts depending on the application will create both relational and multi-dimensional data.

3.7 A Typical Example from the Online Marketing Area

In online marketing, you are very often faced with data marts that are created regularly (once a day) out of a log file data stream or out of a data warehouse counting log file and clickstream data. So the statisticians or other persons with analytical skills can use this data mart as a starting point for their analysis, without investing a lot of time in data preparation. Because for most websites the number of users or unique clients is too big, some of the data marts just include a representative sample of the users or unique clients. Such a data mart can have the structure shown in Figure 3.8 and Figure 3.9.

It is very likely that the data mart will have around 1000 variables if you are recording the clicks for all the different areas of content. There can be a mass of data; the aforementioned example just shows the sum of clicks on a special content during the last 10 slots (slot = day with activity on the web page).

If you know from your domain knowledge that it might be important to find out the averages or the development of the clicks during the last few slots, or the trend, then you have to extend the definition of the information. Figure 3.10 is an example of how it can look just for one theme like 1021 (Fashion/Clothing) from the example before.

This is just an example, but it might illustrate that also for other industry areas, it might be interesting to invest time and maybe money to think about data marts that will be created regularly by the system. If the analytical data miner personnel have to reconstruct the data every time they need the data, it might cost much more and there is no common ground for analyses done by more than one person.

3.8 Unique Data Marts

The large variety of data marts includes two particular types: permanently available data marts and those that have been created through a unique, sometimes complex, analysis.

3.8.1 Permanent Data Marts

This area includes all the data marts that are condensed from history; they are not changing when more data is available. These data marts do not arise from regular data marts; in fact, they only need to be constructed once initially and

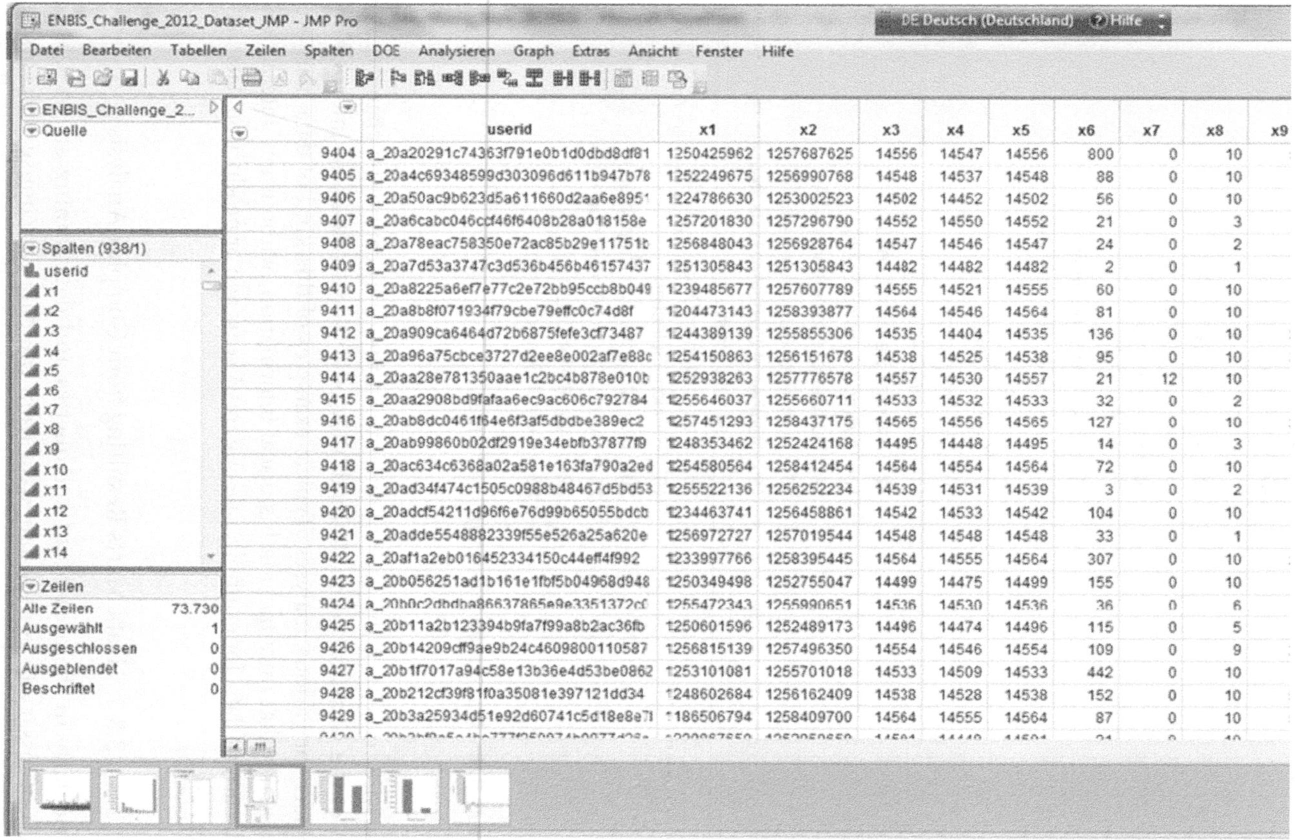

userid	x1	x2	x3	x4	x5	x6	x7	x8	x9
9404 a_20a20291c74363f791e0b1d0d0d8df81	1350425962	1257687625	14556	14547	14556	800	0	10	
9405 a_20a4c69348599d303096d611b947b78	1252249675	1256990768	14548	14537	14548	88	0	10	
9406 a_20a50ac9b623d5a611660d2aa6e895	1224786630	1253002523	14502	14462	14502	56	0	10	
9407 a_20a6cabc046cd46f6408a28a0181580	1257201830	1257296790	14552	14550	14552	21	0	3	
9408 a_20a78eac758350e72ac85b29e11751b	1256848043	1256928764	14547	14546	14547	24	0	2	
9409 a_20a7d53a3747c3d536b456b46157437	1251305843	1251305843	14482	14482	14482	2	0	1	
9410 a_20a8225a6ef7e77c2e72bb95ccb8b049	1239485677	1257607789	14555	14521	14555	60	0	10	
9411 a_20a8b8f071934f79cbe79eff00c7dd8f	1204473143	1258393877	14564	14546	14564	81	0	10	
9412 a_20a909ca646d72b6875fefe3cf73487	1244389139	1255855306	14535	14404	14535	136	0	10	
9413 a_20a96a75cbce3727d2ee8e00d2af7e68c	1254150863	1256151678	14538	14525	14538	95	0	10	
9414 a_20aa28e7813350aae1c2bc4b878e010b	1252938263	1257776578	14557	14530	14557	21	12	10	
9415 a_20aa2908b09fafaa6ec9ac606c792784	1255646037	1255660711	14533	14532	14533	32	0	2	
9416 a_20ab8dc046f6f4e6f3af5dbdbe389ec2	1257451293	1258437175	14565	14556	14565	127	0	10	
9417 a_20ab99860b02df2919e34ebfb3787f9	1248353462	1252424168	14495	14448	14495	14	0	3	
9418 a_20ac634c6368a02a581e163fa790a2ed	1254580564	1258412454	14564	14554	14564	72	0	10	
9419 a_20ad34f474c1505c09889a8467d5bd53	1255522136	1256252234	14539	14531	14539	3	0	2	
9420 a_20adcf542110d9f6e76d99b65055bdcb	1234463741	1256458861	14542	14533	14542	104	0	10	
9421 a_20adde554888233f9f55e526a25a620e	1256972727	1257019544	14548	14548	14548	33	0	1	
9422 a_20af1a2eb0164523341150c44eff4f992	1233997766	1258395445	14564	14555	14564	307	0	10	
9423 a_20b056251ad1b161e1fb9f5b04968d948	1250349498	1252755047	14499	14475	14499	155	0	10	
9424 a_20bb-04bdba86637865aae3351372cf	1256472343	1255990651	14536	14530	14536	36	0	6	
9425 a_20b11a2b1233904b9fa7f99a8b22ac36fb	1250601596	1252489173	14496	14474	14496	115	0	5	
9426 a_20b14209cff9ae9b24c4c609800110587	1258815139	1257496350	14554	14546	14554	109	0	9	
9427 a_20b1f7017a94c58e13b36e4d53be0862	1253101081	1255701018	14533	14509	14533	442	0	10	
9428 a_20b212cf39f81f0a35081e3f7121d34	1248602684	1256162409	14538	14528	14538	152	0	10	
9429 a_20b3a25934d51e92d60741c5d18e8e71	1185506794	1258409700	14564	14555	14564	87	0	10	

Zeilen
Alle Zeilen 73.730
Ausgewählt 0
Ausgeschlossen 0
Ausgeblendet 0
Beschriftet 0

Spalten (938/1)
userid
x1
x2
x3
x4
x5
x6
x7
x8
x9
x10
x11
x12
x13
x14

Figure 3.8 Example data structure.

ID-Variable		Description-Variable
	ID	User/Unique Client
	1	CreationTime
	2	LastSlotCreationTime
	3	CurrentSlotIndex
	4	StartSlotIndex
	5	EndSlotIndex
	6	Clicks with Properties
	7	Clicks without Properties
	8	Slot Count
	9	Session Count
	10	Slot Frequency
	11	Session Frequency
	12	Average Session Length
	13	Average Session Count
	14	Number of Profile Items
	15	Number of Customer Items
	16	Number of Customer Data Items
	17	Has valid Profile
	18	Has Customer Data
	19	Random Selection
	20	Overall Clicks with Properties
	21	Overall Clicks without Properties
	22	Overall Slot Count
	23	Overall Session Count

FIGURE 3.9   Translation list of variable names.

otherwise (except in the case of an error in the data) cannot be changed. As part of the workflow to be defined, they only play a role as an information provider for various analyses. Typical examples of this kind of data mart are data marts that only include historical information, without any interaction with current data, for example, sales figures 3 years ago or production data half a year ago.

3.8.2   Data Marts Resulting from Complex Analysis

Particularly in the area of *ad hoc* queries and data mining analysis, it is often necessary to consolidate data from the data warehouse in terms of different views and contexts. To implement this, there are basically two

General Fashion/Clothing: Profile ID 1021

1021_sum_all	Sum of All Clicks in Lifetime
1021_sum	Sum of all clicks during the last 10 slots
1021_1	Sum of clicks in the last slot (–1)
1021_2	Sum of clicks in the secondlast slot (–2)
1021_3	Sum of clicks in the thirdlast slot (–3)
1021_4	Sum of clicks four times ago (–4)
1021_AVG	Average of clicks all 10 slots
1021_AVG_1	Average of click in the last slot
1021_AVG_2	Average of click in the two last slots (S – 1 and S – 2)
1021_AVG_3	Average of click in the three last slots (S – 1, S – 2 and S – 3)
1021_AVG_4	Average of click in the four last slots (S – 1, S – 2, S – 3 and S – 4)
1021_PROZ	Percent of clicks on the subject (during last 10 slots) compared with all clicks (during last 10 slots)
1021_PROZ_1	Percent of clicks on the subject during last slots compared with all clicks during last slots
1021_PROZ_2	Percent of clicks on the subject during second last slots compared with all clicks during secondlast slots
1021_PROZ_3	Percent of clicks on the subject during third last slots compared with all clicks during third last slots
1021_PROZ_4	Percent of clicks on the subject during fourth last slots compared with all clicks during fourth last slots
1021_TREND	Depending on the chosen trend function, this variable indicates whether the interest in the subject is growing or not

FIGURE 3.10 Example of click information.

possibilities: empower the user or allow the analyst to contact the data warehouse administration with a request to create a data mart according to specified rules or provide the empowered user with an appropriate software solution so that they can create the appropriate data marts by themselves. These kinds of data marts are only done to solve one problem or to carry out a special kind of analysis. We recommend that the user takes a break after finishing the actual analysis and a critical review to see whether the resulting data mart contains variables or ideas that will be fruitful if the data mart becomes one of the regularly implemented data marts.

3.9 DATA MART: DO'S AND DON'TS

There are important issues when creating a data mart. These concern the creation process, the handling of the data mart and the coding/programming aspects. It is worthwhile considering the do's and don'ts for each of these areas when creating a data mart.

3.9.1 Do's and Don'ts for Processes

1. Don't forget the relevant background and domain knowledge.
2. Do use check sums; they should match the numbers you know from your reporting.
3. Do cross-check between different tables and source systems; check that the results fit together and that they represent the relations and averages you have in mind.
4. Don't start the analysis and/or estimation too early.
5. Do define meaningful meta-data.
6. Do prepare and /or transform the data to a shape that suits the methods you plan to use.
7. Do explore the data using easy descriptive analysis and graphical representations.
8. Do carry out peer review; if no second analyst is available, review it yourself the next day.

3.9.2 Do's and Don'ts for Handling

1. Do make sure you can identify and trace every record (by suitable ID variables) at every point so as to enable cross-checking with the original data. Don't ever drop the ID variables even if they are not specifically useful for the analytics.
2. Don't ever lose a relevant case from a dataset; it can happen quite easily, for example, during SQL statements. In particular, cases where 'nothing happened' in one of several tables are likely to disappear during a joint manipulation.
3. Do use meaningful units to categorise continuous variables; the units can be determined based on statistics or business rules.
4. Do check the distributions of variables.
5. Do use meaningful variable names or labels; it is quite handy to show the units (e.g. kg, m, day, euro) in the variable name.

3.9.3   Do's and Don'ts for Coding/Programming

1. Do use options to optimise your dataset (if available in the programming language).
2. Don't forget to check the log information in great detail.
3. Do structure your code.
4. Do use shortcuts and/or macros for frequently used bits of code.
5. Do document the code so that a third person will get a clear picture of what is happening.

In summary, data is the main material for your analysis and decision making. Using the do's and don'ts and the aforementioned guidelines, you should end up with a clear and logical dataset with which to start. The next step is data preparation which involves manipulating the data in more detail.

4

Data Preparation

A Practical Guide to Data Mining for Business and Industry, First Edition.
Andrea Ahlemeyer-Stubbe and Shirley Coleman.
© 2014 John Wiley & Sons, Ltd. Published 2014 by John Wiley & Sons, Ltd.
Companion website: www.wiley.com/go/data_mining

4.1 NECESSITY OF DATA PREPARATION

Having obtained useful data, it now needs to be prepared for analysis. It is not unusual to have the data stored at quite a detailed level in a data warehouse. But to get relevant, reliable and repeatable results out of the analyses, transformation and aggregation of the data is necessary. The type of aggregation has a major impact on the final result. It is unlikely that data mining algorithms will find hidden patterns without prior data preparation. Even if the user doing the data mining is not able to do the transformations and aggregations, it is important for the user to define the necessary steps and make sure someone else does them, for example, colleagues in the IT department. Most of the time, it is also useful to apply domain knowledge in the way the data preparation is done, for example, taking advantage of knowledge of the usual range and type of data items.

4.2 FROM SMALL AND LONG TO SHORT AND WIDE

In most data warehouses, you are faced with data structures that have a central part with multiple associated areas attached in a structure somewhat similar to a snowflake. Let us say that in the centre is quite a big fact-based table where each row represents a detailed piece of information, for example, regarding customer behaviour, or payment attributes, or production details, or advertising and sales campaigns. This centre fact-based table is connected by key variables to other fact-based tables, for example, key variable orders connected to payments, and each of the fact-based tables is surrounded by multi-dimensional tables that give additional information to the keys used in the fact-based table.

Consider the simple example for a manufacturer in Figure 4.1.

If you have a closer look at the 'orders' in the fact table in Figure 4.2, you will notice that the same customer is represented in more than one record. To describe the customer's behaviour and to make him or her suitable for most of the analytical methods, you need to sum the records up somehow. The easiest way might be to sum everything; but information collected yesterday does not always have the same value as that collected more than two years ago. The business may have changed dramatically over the years. Therefore, a weighted sum may be more suitable with some of the data having more importance than other data.

In Figure 4.2, the first column is the row number, the second column is Firmen_Num for company ID and the next column is Kunden_Num for

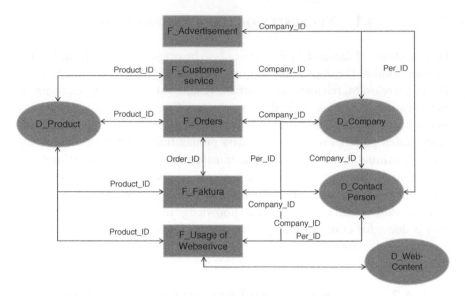

Figure 4.1 Typical connections between fact tables.

customer ID, followed by Auftrags_Num for order, invoice or receipt number; Bestell_Datum for order date; Order_sta for order status; and then two variables to identify which advertisement generated the order. The final two columns are Betrag_Auftrag for order value and Zahl_Art for how it is paid.

In this excerpt of an order fact table, notice that, for example, rows 84040–84047 have the same customer ID and there are seven different order IDs. The oldest order dates back to the year 2011 and the youngest to the year 2013. This example shows that a single customer can appear in several rows of the order fact table, each row corresponding to a different value for one of the other variables: product, order and date.

If the subject of our analysis is the customer, we should plan to end up with just one row for the customer, for example, Kunden_num 374172. If the subject of our analysis is the company, we should plan to end up with just one row for the company.

If we carry out a simple screening of all the rows, we might come up with the following in Figure 4.3 with either company as subject or customer as subject.

This is a very easy way to summarise the different records for one subject, but it does not use all the information included in the original data. Regarding reading and simplification, we can change the format of description, and we are just rearranging what is already there. What else can be found in the information at the customer level? Consider defining a particular day, 15 January 2014, and summarising the information as shown in Figure 4.4.

Row		Firmen_Num	Kunden_Num	Auftrags_Num	Bestell_Datum	Order_Sta	WA_Ident	WA_Stufe	Betrag_Auftrag	Zahl_Art
84040		13	3741729	6935355	22DEC2013	25	B0FMA0159	3	49.40	1
84041		13	3741729	7441253	11APR2011	25	B14SA0000	5	79.40	1
84042		13	3741729	7818626	25JUL2011	25	B13MA0189	9	59.40	1
84043	Customer A	13	3741729	8511391	06FEB2012	25	B21S1FRE	3	78.00	2
84044		13	3741729	9459182	27OCT2012	25	B21SA0KTR	4	78.00	2
84045		13	3741729	9472983	28OCT2012	19	B21SA0WGS	3	0.00	1
84046		13	3741729	9779569	14JAN2013	25	B24MA0181	3	0.00	2
84047		13	3741729	9838150	31JAN2013	25	B24MA0181	3	79.70	2
84048		13	3741737	1022039	08DEC2003	25	913	34	378.20	1
84049		13	3741737	1296776	10JUN2004	25	919	2	179.40	1
84050	Customer B	13	3741737	1296776	10JUN2004	25	919	2	0.00	1
84051		13	3741737	1486276	08SEP2004	25	B41SA035N	1	118.80	1
84052		13	3742229	1015485	04DEC2003	25	043	448	118.80	1
84053		13	3742229	1118773	07FEB2004	25	100	6	302.30	1
84054		13	3742229	1186103	25MAR2004	25	116	18	118.80	1
84055		13	3742229	1215443	14APR2004	25	120	13	118.80	1
84056		13	3742229	1667954	03NOV2004	26	B41MA0008	13	119.96	1
84057	Customer C	13	3742229	1668151	03NOV2004	26	B44MA0003	1	0.00	1
84058		13	3742229	1668508	03NOV2004	25	B44MA0003	1	118.80	1
84059		13	3742229	1678512	08NOV2004	25	B4CSA0000	4	261.13	1
84060		13	3742229	1678512	08NOV2004	25	B4CSA0000	4	69.70	1
84061		13	3742229	2277637	03NOV2005	19	B44MA0003	1	118.80	1
84062		13	3742229	4470107	18MAY2009	25	B93MA0148	8	29.70	1
84063		13	3742229	5031499	19NOV2009	25	B91MA0164	1	197.60	1
84064		13	3742229	5445184	10MAR2013	25	B04MA0148	1	76.40	1
84065		13	3742229	5893738	29JUN2013	25	B01MA0268	2	0.00	1

FIGURE 4.2   An example of an order fact table.

Person as subject:

Kunden_num	Num of Orders	Revenue Total	Num of Distinct Adverts
374172	8	423.9	7

FIGURE 4.3   Person as subject – easy.

Kunden_num	Num of Orders Total	Revenue Total	Num of Distinct Adverts	Revenue_ 30 Days	Revenue_ 180 Days	Revenue_ 365 Days	Num of Orders 365 Days	Num of Orders Status 19
374172	8	423.9	7	49.4	49.4	129.1	3	1

FIGURE 4.4   Person as subject – more information.

Sometimes, especially if you are faced with thousands of different products, it makes more sense to choose product groups or colours or some other material information to define the variables. Note that in the end, the data mart should contain information from all relevant fact tables like in the manufacturer example which included advertisement, online information and customer service. This ensures that revenue information, for example, is available in the data mart as well as the click count of online activity as well as numbers of marketing activities.

This example should give an impression and inspiration. There is no overall master plan for carrying out data mining on the available data marts, but it may help you to decide to create new variables if you try to see and cumulate the information under different viewpoints, even if you know that in a statistical sense, it is quite likely that some of the variables are highly correlated. Depending on the domain, you can create hundreds of new variables from a slim fact table (i.e. a table with only a few columns) or a set of connected fact tables to describe each case or subject (mostly cases are customers).

The only statistical reason that might limit your number of variables is the fact that for most of the analytical methods, the number of records or cases in the analytical dataset should be bigger than the number of relevant variables. If it seems that you will be faced with the problem of too many variables and not enough data, follow a two-step strategy:

1. Create all variables that come to your mind or are common domain knowledge.
2. Use feature reduction methods that help you to find the most relevant variables for your actual problem.

The relevant variable set may vary from analyses to analyses, but note that it is important to recheck that similar problems should have similar variable sets. In case you are faced with the situation that you know from your business (domain) knowledge that two problems are highly correlated but the outcome of your feature reduction is totally different, you must dig deeper to solve the problem or be totally sure of the results. Otherwise, you run the risk of creating unstable models, or you may lose a brand new business rule that might have arisen as part of the outcome.

4.3   TRANSFORMATION OF VARIABLES

Recall that in the example earlier, the data in the data mart may have different scales and sometimes even different measurement levels. For most of the analysis, it is an advantage if the variables have comparable scales and similar measurement levels. Not all measurement levels are suitable for every kind of method without additional transformation. Figure 4.5 gives an overview of the interaction between measurement level and data mining method.

Note that the descriptions of most of the common data mining software tools give the impression that every method can be used with any measurement level. This is only half of the truth. What it really means is that the tool includes automatic methods to transform the data to a suitable level; a well-known example is the transformation from nominal data to several new binary variables, also known as indicator variables. For example, if the original variable consists of colours like red, blue and green, it will be transformed to three variables – one for red, with a value of 1 if the original variable is red and value of 0 if not;

Method	Measurement level			
	Metric/interval	Ordinal	Binary	Nominal
Decision Tree	X	X	X	X
Neural Network	X	X	X	
Regression	X	X	X	X
Support Vector Machines	X	X	X	
Clustering	X			
SOM's	X	X	X	
Association Rules		X	X	X
Sequence Analyses	X	X	X	X
CBR	X	X	X	X

FIGURE 4.5   Interaction between measurement level and data mining method.

one for blue; and one for green – each following the same pattern. In most cases, the automatic transformation is fine and suits the need, but there are cases where it might not be the best solution. For example, it may be that an ordinal scale is really more appropriate such as white = 1, grey = 2 and black = 3. In this case, the required transformation has to be done manually. This kind of transformation and the associated algorithm must be defined using your domain knowledge. For example, the first model constructed using the automatic transformation might miss out rules that you might have expected based on your business knowledge. If these missed rules are related in some way to the critical variable, you would be well advised to do the relevant transformation by hand and redesign the model and then compare the results.

Depending on the chosen or planned data mining method, different scales might have a big influence on model stability and results. Consider these challenges:

- Missing data
- Outliers
- Data with totally different scales
- Data with totally different distributions

4.4 MISSING DATA AND IMPUTATION STRATEGIES

Based on the fact that data for the majority of data mining projects is observational data collected from existing processes as opposed to well-designed experiments, missing data is a common problem. To fix it, we have to distinguish between 'real missing data' and 'not stored information'. Typical examples for real missing data are missed birthday or age information in marketing problems or temperature or moisture measurements in manufacturing datasets. Real missing data occurs in datasets generated from situations where the information itself for sure exists in real life, but for unknown reasons, the information is not stored in the dataset. For example, every customer actually does have an individual age, and he or she is actually born on a particular date even if that data is missing. The data may be missing because of an explicit wish of the customer not to share the information or because some process or other for obtaining the information has not worked correctly. Missing information on temperature or moisture or such like is very dependent on process errors or technical errors.

If real missing data is detected, then it can be replaced by estimation using imputation strategies. The estimation method differs depending on the business context and on other existing information. Estimation can be by replacement using the mean or median (calculated over all data or just a relevant subset) or

Firstname	Salutation	Sexcode
Clemens	Herr	1
Clement	Herr	1
Clement Keit	Herr	1
Clementia	Frau	2
Clementine	Frau	2
Clementino	Herr	1
Clements	Herr	1
Clementus	Herr	1
Clemetina	Frau	2
Clemm	Herr	1
Cleopa	Frau	2
Cleopha	Frau	2
Clere	Frau	2
Cletus	Herr	1
Cliff	Herr	1
Clio	Frau	2
Clito	Herr	1
Clive	Herr	1
Clivia	Frau	2
Clivio	Herr	1
Clodhilde	Frau	2
Clothilde	Frau	2
Clothilde J.	Frau	2
Clotilde	Frau	2

FIGURE 4.6   Example of a look-up table.

may be by a more complex method such as using a regression equation or a time series analysis. Cases with missing variables are only rarely excluded.

An alternative way to replace real missing values is to use third-party knowledge. For example, a look-up table can be used to deduce gender from first names, like Mary = female and John = male. If such a table is not available, then you have to construct one yourself based on the full dataset that you have. Make sure that you use the same coding, for example, if your full dataset uses 1 = male, then you should create a look-up table with 1 = male as well. To combine the dataset containing missing values with the look-up table, you should merge both tables using the first names as key. You can also create a look-up table from the salutation (i.e. Mr(Herr) or Ms(Frau)) (see Figure 4.6).

Missing data that represents 'not stored information' is very different to the cases of real missing data; they have to be replaced by zero or another value that represents the business knowledge behind it. These cases happen because most company databases just store things that happened and not things that have not happened. An example is that customers 'who buy' create footprints in the database; that fact is stored in several tables in which you will find the purchase date, amount, value, products, the way they are ordered, paid, delivered and so on. For those customers 'who did not buy', you will find nothing in the company databases at that time. But for some analyses, it is important to represent the 'not buying' fact as well in the dataset that should be used for data mining, especially if 'not doing something' is done by the majority of people.

Instead of representing the 'not happened case' by zero, you can also count the number of days since the last time something happened. Note that this kind of missing data has to be replaced with a value indicating the business meaning, but in addition it should fit with the preferred analytical method. For example, every 'no buy' can be represented by zero if the variable itself contains values (money) or amount (pieces). If it contains 'days since', then you cannot represent the missing data with zero because that would wrongly lead you to the interpretation that something happened just recently. Here, it may be better to use the number of days since the last known activity (e.g. subscribe for an email newsletter) for estimation or other similar substitutes that correspond with your business rules.

Consider the following example:

Customers: A and B bought in the last six months; C and D did not buy in the last six months.

Today's date: 10.12.2011

Customer A: entry 06.06.2000	Purchases: 04.12.2011	Product 1234	4 999 €
Customer B: entry 28.04.2001	Purchases: 07.12.2011	Product 1234	4 999 €
Customer B: entry 28.04.2001	Purchases: 15.11.2011	Product 5678	14 999 €
Customer C: entry 23.01.2007	Purchases: 23.01.2007	Product 1234	3 999 €
Customer A: entry 06.06.2000	Purchases: 06.06.2000	Product 458	6 999 €
Customer D: entry 06.06.2000	No purchases at all		

Figure 4.7 shows part of a data mart with missing values.

Figure 4.8 shows part of a data mart with replacement for missing values.

Note that if a customer pays by cash, there is no automatic record of their details unless they are willing to divulge them or they can be persuaded to use a loyalty card.

Customer ID	Overall Purchase Value Last 7 Days	Overall Purchase Value Last 30 Days	Days Since Last Purchase	P1234 Purchase Value Last 7 Days	P1234 Purchase Value Last 30 Days	P1234 Since Last Purchase
A	49.99	49.99	6	49.99	49.99	6
B	49.99	199.98	3	49.99	49.99	3
C	.	.	1782	.	.	1782
D	.	.	.	.	.	.

FIGURE 4.7 Part of a data mart with missing values.

Customer ID	Overall Purchases Value Last 7 Days	Overall Purchases Value Last 30 Days	Days Since Last Purchases	P1234 Purchases Value Last 7 Days	P1234 Purchases Value Last 30 Days	P1234 Days Since Last Purchases
A	49.99	49.99	6	49.99	49.99	6
B	49.99	199.98	3	49.99	49.99	3
C	0	0	1782	0	0	1782
D	0	0	4204	0	0	4204

FIGURE 4.8 Part of a data mart with replacement for missing values.

4.5 OUTLIERS

Outliers are unusual values that show up as very different to other values in the dataset. Outliers can arise by process errors or through very rare customer behaviour or other unusual occurrences. If it is a process error, you can handle the outlier in a manner comparable to a 'real missing' value, or if you have enough cases, you can reject the case that includes the outlier. If you are sure that it is not a process error, then it is quite common to use techniques such as standardisation or binning or the quantile method to mitigate the effect of the outlier. These methods are described in the next sections. Otherwise, it may be sensible to compare the results of analysis with and without the outliers and if there are marked differences to report on both sets of results.

4.6 DEALING WITH THE VAGARIES OF DATA

4.6.1 Distributions

For most of the data mining methods, there are no constraints or only minor constraints about the distribution of the variables. Regression techniques require that target data follows the Normal distribution after allowing for the explanatory variables in the regression. Principal components analysis and factor analysis also require the variables to be Normal.

4.6.2 Tests for Normality

A reasonable test for Normality is to look at a histogram of the data. If the histogram is approximately bell shaped and symmetrical, then the data can be assumed to be approximately Normal. More precise tests can be used, for example, the Kolmogorov–Smirnov test (see statistical texts in Bibliography). However, when there are large quantities of data, any deviation from Normality will be flagged up as statistically significant. Hence, a visual appraisal is usually satisfactory. Tests for Normality are important as they show up features in the data. A common reason for lack of Normality is that there are really two or more sub-groups within the data. It is important to recognise these, for example, customers from one location may have totally different shopping habits from those from another location. Separate analyses may be appropriate for these sub-groups, or the test for Normality must be carried out after fitting an explanatory variable included to represent location.

Many data items are approximately Normally distributed when they are representative of a distinct subset of the dataset. However, some measures are intrinsically non-Normal. For example, the lifetime of customers is unlikely to be Normal as no one can have a lifetime less than zero, yet some loyal customers can stay with the company for many years; therefore, the distribution will be asymmetrical and can only extend to large values on the positive side. In this case, the histogram will be skewed to the right. Normalisation and ranking or other transformations like a Box–Cox power transformation (see succeeding text) can help to make the data approximately Normal. For positively skewed data, a log transformation is effective for making the data more symmetrical. For left skewed, other transformations have to be tried.

4.6.3 Data with Totally Different Scales

Data from real processes sometimes have totally different scales, for example, dates have a different scale to the number of items sold, to the money that has been paid and to simple binary data like payment with credit card or not. In this case, it is necessary to think about methods that equalise the scales. Again, you can solve this problem with standardisation/normalisation, ranking or binning. These issues are dealt with in the succeeding text.

4.7 Adjusting the Data Distributions

4.7.1 Standardisation and Normalisation

Standardisation usually refers to the process of scaling data so that it has a zero mean and unit standard deviation. This may be carried out to ensure that two or more sets of data have the same scale. The usual reason for carrying out standardisation is to be able to compare the variables.

Standardisation is usually intended to make data Normal, in which case it can be considered to be normalisation. Normalisation is carried out when you need a Normally distributed variable because you want to use an estimation method that requires Normality or at least variables with comparable scales. Note that if you transform using a standard normalisation, you will end up with a variable having a mean equal to 0 and a standard deviation equal to 1. Normalisation is only relevant for continuous data. It is a very common procedure and exists as a function in nearly all data mining or statistical tool sets.

Standardisation and normalisation may make the interpretation of the resulting models and the transformed variables more difficult especially for non-statisticians. If you use normalisation, also available in the common statistical and data mining tools, the transformation will deliver a meaningful and comparable mean on one hand and variable scale in a specific range on the other.

4.7.2 Ranking

Ranking is a possible way to transform the distribution of a variable and is also a good opportunity to deal with business-related outliers. Ranking is very easy to do; some data mining tools provide a procedure for it. Otherwise, you can do it yourself by sorting the variable by size and giving a rank order number for each row. Whether rank one is the highest or the lowest value depends on the domain context. If the new variable should follow the Normal distribution, an additional simple transformation is required.

4.7.3 Box–Cox Transformation

Alternatively, a Box–Cox transformation is done which results in the transformed values being a power of the original. Note that zero remains zero. If the data is unstable so that there are lots of business-related changes, then the Box–Cox method can be too sensitive as it may change with each change in the data; in other words, it is like an over-fitting problem, and it gives the impression of being more exact than is justified. The Box–Cox transformation lacks robustness in our context. However, it can be a useful method if your data is stable.

4.8 BINNING

To stabilise and improve the data mining models, it is preferable to classify continuous variables, such as turnover, amount or purchasing days into different levels. Using such a classification, it is possible to stress more strongly differences between levels that are important from the business point of view.

This is particularly the case with variables that are conceptually non-linear. For example, it is important to know that a buyer belongs to the 10% of best buyers, but the numerical distance between turnovers of, say, 2516 € and 5035 € is less important. In contrast, 1 € is rather like 0 € from a mathematical and statistical point of view; however, from a business point of view, 1 € tells us that the person has actually made a purchase, however small, and is therefore a better prospect than someone who has made no purchase at all.

The binning technique can be used for all kinds of data. There are two ways to do the binning: in consideration of business rules and in consideration of statistics and analytics.

The binning of age is an example in consideration of business rules. For a lot of analysis, age groups like 25–30 years or 45–50 years are used. The basic idea behind it is to group the variable values to new values to reduce the amount of values. In addition to immediate business considerations, it may be advantageous to align the binning with groups used by the National Statistics Institutes in official statistics; hence, age groups such as 46–50, 51–55, 56–60 may be preferable so that we can compare our results with information and open data that are publically available.

The method of quantiles described in the succeeding text is an important way to do an analytical binning on continuous or ordinal variables. It is even useful for ordinal variables when there are fewer than six levels because the binning puts all the data on the same level of granulation.

An analytical binning is also possible for nominal variables; the same algorithms can be employed as are used to build up a decision tree in relation to the target variable.

4.8.1 Bucket Method

The bucket method refers to putting data into pre-defined categories or buckets. It is a special form of binning that can be carried out by hand to fulfil the purpose of transformation. An example is shown in Figure 4.9.

Note that the chosen transformation must coincide with the unwritten rules of the business, for example, that customers from rural areas are grouped together. Some variables, such as ordered postcodes, do not give a good grouping. In the case that single variables are filled with numbers representing codes, you have to do your grouping for this variable by hand or create dummy variables out of it.

The bucket method has to be done for each variable separately based on business knowledge or given categories. We recommend thinking carefully about whether the bucket method is necessary and gives added value; otherwise, it is probably better to use an algorithm as the benefit gained from a more tailored approach is not usually worth the additional manual intervention.

4.8.2 Analytical Binning for Nominal Variables

Binning is carried out in various data mining algorithms if there are a large number of categories for nominal data. For example, binning is carried out on variables used to build up decision trees in relation to the target variable. The binning algorithms search for meaningful partitions of the variables. Further details can be found in standard data mining texts in the Bibliography.

4.8.3 Quantiles

Based on our experience, the quantile method is a very easy and robust method. To transform data to quantiles, we will first order the data according to its values and then divide the rows into a number of quantiles. Often a useful choice is to use six quantiles, or sextiles; this is a compromise between too much variation in value and too little. Our practical experience is that it is more successful to use six sextiles instead of the classic four quartiles often used in social science which give the five quantile borders as minimum, lower quartile, median, upper quartile and maximum.

For example, if there are 250 000 cases (rows) and we want to convert to sextiles, then there will be 250 000/6 values in each sextile. We note the values of the sextile borders and use an 'if then else' construction to transform the real data into six classes with values between 0 and 5. However, there are often a lot of tied values in practical data mining data in which case not all values from 0 to 5 may emerge. For example, if the class border of the 1st sextile is zero, then the transformed variable is zero. If the class border of the 2nd sextile is zero, then the transformed variable is also zero. The first sextile border which is non-zero is transformed to the next order value of the sextile, so if the first non-zero border is for the 4th sextile, then the transformed value is 4. So transformed values 1, 2 and 3 have been omitted. If the next sextile value is the same, then the transformed value is still 4. If the value at the next sextile border is higher, then the next sextile value is awarded.

Note that typically we can have a lot of zeros for some variables, and the first few quantiles will be zero with only the last quantile taking the final value of 5. In this case, instead of having a quantile transformation with values 0 and 5, we might prefer to consider the transformation to a binary 0/1 variable instead. However, we may prefer to impose a consistent method for all the many variables rather than using a different method for each variable.

4.8.4 Binning in Practice

Classification of variables can be realised in two ways:

- Commonly, data mining tools allow the binning procedure: first of all, you have to evaluate the sextile border for every single variable and then deal with every variable individually by hand. The advantage of a very individual classification, however, is bought with the disadvantage of the very high lead time.
- Alternatively, you can use coding facilities, for example, with a SAS standard programme to classify every variable more or less automatically. The idea behind this is that the sextiles into which the variables are divided are fixed by the programme which then gives the class borders.

To make it clear, we will give two examples using SAS Code but described without assuming any experience in SAS coding.

Example

Number of pieces ordered for industry 21 for 3 business year quarters

Scalar variable	Sextile	Border
M_BR_0021_3ya	The 16.67 percentile, M_BR_0021_3	0
M_BR_0021_3yb	The 33.33 percentile, M_BR_0021_3	0
M_BR_0021_3yc	The 50.00 percentile, M_BR_0021_3	0

Scalar variable	Sextile	Border
M_BR_0021_3yd	The 66.67 percentile, M_BR_0021_3	0
M_BR_0021_3ye	The 83.33 percentile, M_BR_0021_3	1
M_BR_0021_3yf	The 100.00 percentile, M_BR_0021_3	225

The procedure is carried out in the following classification rules:

SAS_CODE:

```
select;
when (M_BR_0021_3 <= M_BR_0021_3ya) CM_BR_0021_3 = 0;
when (M_BR_0021_3 <= M_BR_0021_3yb) CM_BR_0021_3 = 1;
when (M_BR_0021_3 <= M_BR_0021_3yc) CM_BR_0021_3 = 2;
when (M_BR_0021_3 <= M_BR_0021_3yd) CM_BR_0021_3 = 3;
when (M_BR_0021_3 <= M_BR_0021_3ye) CM_BR_0021_3 = 4;
otherwise CM_BR_0021_3 = 5;
;
```

In summary, the rule for transforming M_BR_0021_3 to CM_BR_0021_3 is:

M_BR_0021_3 <= 0 => Classification Variables CM_BR_0021_3 the value = 0
M_BR_0021_3 = 1 => Classification Variables CM_BR_0021_3 the value = 4
M_BR_0021_3 > 1 => Classification Variables CM_BR_0021_3 the value = 5

Using this coding, the distance between the values 0 'has not shopped' and 1 'exactly 1 piece has been bought' has been artificially increased to be more meaningful in terms of the business.

Now, consider variables that are well defined, as, for example, the annual turnover from a random sample of customers who have joined the company in the last 15 months. Here, the distance between customers with turnovers of 906.55 € and 45 149.67 € is diminished because all customers with more than 906.55 € annual turnover belong to the best 16%. This avoids the potential distortion due to the extremely large value of 45 149.67 €.

The procedure is carried out in the classification rules shown in Figure 4.10.

The standard programme selects the relevant variables from the meta-data with the help of a macro (i.e. a user-provided set of instructions). Then the Procedure (Proc) Univariate is carried out for these variables twice: firstly, for the general analysis of the single variables and, secondly, to determine the borders of the sextiles. The results, that is, the borders, pass into a programme step which

Variable
U_GES_12M (total revenue last 12 Month)
Range from 0 to 45,149.65 Euro

Binning Example I :
Bin 1 :   U_GES_12M = 0
Bin 2 :   0 < U_GES_12M <= 50
Bin 3 :   50 < U_GES_12M <= 100
Bin 4 :   100 < U_GES_12M <= 500
Bin 5 :   500 < U_GES_12M <= 1000
Bin 6 : 1000 < U_GES_12M <= 5000
Bin 7 : 5000 < U_GES_12M

Binning Example II:
Bin 1 :   U_GES_12M <= 100
Bin 2 :   100 < U_GES_12M <= 1000
Bin 3 : 1000 < U_GES_12M

FIGURE 4.9   Dealing with binning.

Variable	Sextile	Border
U_GES_12Mya	the 16.6700 percentile, U_GES_12M	73.43
U_GES_12Myb	the 33.3300 percentile, U_GES_12M	176.80
U_GES_12Myc	the 50.0000 percentile, U_GES_12M	310.34
U_GES_12Myd	the 66.6700 percentile, U_GES_12M	516.51
U_GES_12Mye	the 83.3300 percentile, U_GES_12M	906.54
U_GES_12Myf	the 100.0 percentile, U_GES_12M	45,149.65

FIGURE 4.10   Dealing with sextiles.

The procedure is carried out in the following classification rules:

```
SAS_CODE:
select;
when (U_GES_12M <= 73.43)   CU_GES_12M = 0;
when (U_GES_12M <= 176.8)   CU_GES_12M = 1;
when (U_GES_12M <= 310.34) CU_GES_12M = 2;
when (U_GES_12M <= 516.51) CU_GES_12M = 3;
when (U_GES_12M <= 906.54) CU_GES_12M = 4;
otherwise CU_GES_12M = 5;
;
```

Rule:
U_GES_12M <= 73.43   => Class. Variables CU_GES_12M the value = 0
73.43 < U_GES_12M <= 176.8   => Class. Variables CU_GES_12M the value = 1
176.8 < U_GES_12M <= 310.34   => Class. Variables CU_GES_12M the value = 2
310.34 < U_GES_12M <= 516.51 => Class. Variables CU_GES_12M the value = 3
516.51 < U_GES_12M <= 906.54 => Class. Variables CU_GES_12M the value = 4
U_GES_12M > 906.54   => Class. Variables CU_GES_12M the value = 5

FIGURE 4.11   Classification rules.

independently generates a programme code and exports it, with the eventual outcome that we obtain a file with the classified variables (see Figure 4.11).

After the data preparation and transfers are done, you can do the planned further analytical steps. If you like to solve prediction problems, it may help if you will have a feature reduction step before you conclude with the predictions.

4.9   Timing Considerations

Considerable time needs to be allowed for data preparation. As can be seen from the sections earlier, there are many different things to be aware of and steps to be taken. In common with most projects, time spent on preparation is seldom wasted.

The time needed for data preparation depends not only on the data but also on the experience of the practitioner. A more experienced data miner will need less time than a novice. It requires programming skills and familiarity with data preparation tools. Coded data can usually be prepared more quickly. Data preparation can take one to two days up to two to three weeks. In case the data structure is not clear, never estimate less than a week preparation time. Thereafter, the data may be in use for a long period.

Another issue is whether you have access to the data and whether you have all the necessary authorisation or have to ask other people to do things for you. It is quicker if you can do it yourself providing you are experienced. If you are not experienced, it may be better to have someone else do the data preparation for you.

Data is dynamic and frequently updated. In some cases, the data needs to be re-prepared weekly before it can be used.

4.10   Operational Issues

It is vital that the analyst realises the importance of understanding the definitions that the company uses. It should be noted that often the company may not have agreement on this within themselves. Whatever definitions the analyst uses, their results have to be compatible with those used in the marketing and customer relationship management reports. For example, does average age include missing values as zero or not, and are missing values included in the analytics or not? If zeros are included, a quantity like average age will end up being much lower than it is known to be. One way to check such important details is to map the analytical results to those in other business reports.

5

Analytics

Analytics

A Practical Guide to Data Mining for Business and Industry, First Edition.
Andrea Ahlemeyer-Stubbe and Shirley Coleman.
© 2014 John Wiley & Sons, Ltd. Published 2014 by John Wiley & Sons, Ltd.
Companion website: www.wiley.com/go/data_mining

5.1   Introduction

Data needs to be explored and analysed, and decisions need to be made. These activities are sometimes referred to as descriptive analytics and predictive analytics. Here, we include details of carrying out comparative tests and cross tabulations and consider how to detect correlations and patterns in the data that can be useful in selecting variables to build a model.

In data mining, there are special issues when testing features of the data because of the size of the datasets; statistical significance tests have to be interpreted differently. There are also issues when working with subsets sampled from the data, how the subsets are chosen and how we can check whether the subsets are representative of the whole dataset. It is quite common to develop a number of models using different methods and different subsets of the data. The models can then be compared in terms of their fit and their stability and also in terms of their business relevance. This chapter includes ways to choose the model which has the best quality and is the most stable.

In the descriptive analytics approach, we can explore the data by looking at summary statistics such as measures of magnitude and dispersion for continuous variables or proportions for categorical variables, and we can look at how one variable relates to another variable by carrying out cross tabulations. Cross tabulations may be in the form of a single variable such as 'gender' in the rows against the target such as 'buy' or 'not buy' in the columns of the table and can lead to tests of importance of the single variable based on chi-square tests of independence (generated, e.g. in SAS with a PROC Univariate and PROC Freq). If the target is independent of the single variable, then the single variable is not likely to be very important. If the target is related to the single variable, then it is potentially important in helping us to achieve the target.

In the predictive analytics approach, we can carry out further analyses like regressions, decision trees, neural networks, etc. These analyses can be carried out with simple random samples of the data (e.g. a 20% sample) as well as with a stratified random sample. The simple random sample is preferred if there are plenty of target customers for the size of dataset. For example, 1% of people may be churn customers, so a 20% random sample will contain 0.2% of churn customers, and provided the dataset is large enough, this ensures that you are working with enough of the target people (i.e. at least 100 people, which can be achieved if the dataset has more than $100/0.2\% = 50\,000$ cases). However, if the population proportion of target customers is very small or the population is small, then a simple random sample may not include any target customers, and a stratified random sample is preferred.

Both of these approaches can be tackled using suitable software like SAS Enterprise Miner or JMP. Firstly, we can use the 'Explore'-related functions to carry out the analysis of measures of magnitude and dispersion or cross tabulation. We can then use the functions around 'Data partition' to generate at least one training sample (training) and a test sample (validation). Then, the system develops the models on the training sample and checks them on the test sample automatically; models can be calculated using different procedures with different algorithms and settings and are compared with each other. The quality and stability of the models determine which model is chosen.

5.2 BASIS OF STATISTICAL TESTS

5.2.1 Hypothesis Tests and *P* Values

There are many different types of statistical tests:

- To compare sample statistics with population values to assess the quality of the sample
- To test for independence between variables and target to find variables that are related to the target and therefore likely to be useful for prediction

The basis of statistical tests is that a null hypothesis is stated and a corresponding alternative hypothesis is specified; then, a test statistic is calculated. The size of the test statistic is compared to standard tables, and on that basis, it is decided whether the null hypothesis should be rejected in favour of the alternative hypothesis or whether the null hypothesis should be accepted. Null hypothesis is written as H0, and the alternative hypothesis is written as H1.

	H0 rejected	H0 accepted
H0 false	Right decision	Type 2 error
H0 true	Type 1 error	Right decision

FIGURE 5.1   Outcome of a hypothesis test.

For example:

H0: The variables are independent of each other.
H1: The variables are not independent of each other.

The aim of the testing is usually to reject the null hypothesis as this is usually the more useful or interesting outcome. There are two possible outcomes for a hypothesis test: accept or reject the null hypothesis. There are also two possibilities for the underlying truth of the situation. Recall that the hypothesis test is carried out on a sample of data with a view to determining the truth about the population. There are, therefore, four possible situations which can result from a hypothesis test. These four situations are shown in Figure 5.1.

The probability of rejecting the null hypothesis, although it is actually true, is called the type I error; this is often represented by P, p or α. The probability of accepting the null hypothesis, although it is actually wrong, is called the type II error, represented by β.

The usual significance borders are:

$P <= 0.1$ tending towards significance (10% significance level)
$P <= 0.05$ significant (5% significance level)
$P <= 0.01$ very significant (1% significance level)
$P <= 0.001$ very highly significant (0.1% significance level)

Significance tests result in a statistic like the t value or a chi-square value and a corresponding p value which is the probability of obtaining such a value for the statistic if the null hypothesis is true. The p value is compared to the significance borders (see Figure 5.2).

In special cases, it can also make sense to accept significance borders of 15% or 20% and still include the variable as possibly important. One such case is in decision trees which determine the branches on the basis of chi-square tests.

```
xNK_AKA_YYT1
Frequency
Percent
Row Pct          Target
Col Pct            0  |     1  |  Total
```

		Target 0	1	Total
0		1976	1785	3761
		49.40	44.63	94.03
		52.54	47.46	
		98.80	89.25	
5		24	215	239
		0.60	5.38	5.98
		10.04	89.96	
		1.20	10.75	
Total		2000	2000	4000
		50.00	50.00	100.00

Statistic	DF	Value	Prob
Chi-Square	1	162.3400	< .0001
Phi Coefficient		0.2015	
Contingency Coefficient		0.1975	
Cramer's V		0.2015	

FIGURE 5.2 Example of a significance test.

There, it often makes sense to raise the significance border to ensure that variables that are only slightly significant are still considered. This is because although the variable is not directly dependent, it may be dependent in an interrelated way, that is, in an interaction.

5.2.2 Tolerance Intervals

Data is subject to random variation. The sources of the random variation include random choice of purchase, timing considerations, recording errors, different types of people. If we want to summarise the age of our customers, the mean value just gives us a measure of magnitude, for example, 45 years. It makes a difference to our interpretation of the age of customers whether all customers were exactly 45 years old or whether half were 30 and half were 60 years old. The variation in a quantity such as age of customer is summarised by the standard deviation. The likely range of the variable is given by a tolerance interval. The likelihood of values being in the tolerance interval depends on the confidence level chosen.

5.2.3   Standard Errors and Confidence Intervals

Often, a sample of customers is analysed to estimate the behaviour of the whole population. Larger samples lead to more precise estimates. The variation in an estimate is summarised by the standard error.

The likely value of a population parameter, such as the mean, is given by the sample mean value which is a point estimate. The standard error of the sample mean is smaller for larger samples and is calculated as the sample standard deviation divided by √sample size. Based on a sample of n customers, with mean age m and standard deviation s, the standard error is $se = s/\sqrt{n}$.

The 95% confidence interval for the population mean value is

95% confidence interval: $m - (1.96 \times se)$ to $m + (1.96 \times se)$

Confidence intervals are calculated whenever parameters are estimated from statistical models.

5.3   SAMPLING

5.3.1   Methods

Data examination should be carried out on the full dataset. Data preparation should also be done on the full dataset if deployment will apply to the whole dataset. If deployment is not intended for the whole dataset, for example, if we are just looking for patterns and to understand the data, then we can carry out our analysis on a subset or sample.

When we come to build models, then we can use samples which may be anywhere from 5 000 to 100 000 or more. However, with modern computer power, we can carry out much of our analysis using the whole dataset even if it has many millions of records.

We need to understand the structure of the data to carry out meaningful partition when we want to make sensible predictions.

First of all, we will describe our data in terms of the target variables of interest. We may find, for example, data such as in Figure 5.3.

If we take a random sample from this dataset, we will get mostly non-churners. So we prefer to take a stratified random sample to give a good proportion of churners and non-churners. In some cases, we will take all of the churners and only take a random sample of the non-churners. We can use the provision in JMP or any other software to select the random sample. We may take equal numbers from each of the churners and non-churners.

Target	Yes	No	%yes	%no
Churn	3 987	3 544 777	0.1%	99.9%
Buy (next four weeks)	64 546	356 982	15.3%	84.7%
Reactivation	4 546	698 277	0.6%	99.4%
Click	932	13 054 678	0.0071%	99.9%

FIGURE 5.3   Target data.

5.3.2   Sample Sizes

Determining the appropriate sample size is an important issue in statistical analysis. However, in data mining, we have a different viewpoint. The population size of data may be enormous, and 10 million cases are not uncommon. Indeed, 'big data' is becoming more mainstream to data mining analysis, and so the size of the dataset is increasing all the time. Therefore, it is not problematic to select a large sample; however, even in data mining, the sample quality still remains an issue. The sample should be representative of the population, at least as regards key variables.

One of the main constraints is that the sample size must exceed the number of variables used in the models. After transformation, classification and binning, there can be thousands of variables, and 7–8000 variables are not uncommon. So, if there is enough data, sample sizes bigger than 10 000 are not uncommon in data mining.

The sample size depends on the original size of the dataset; for example, if the population is 10 million or more, then a sample of 10 000 is not enough as it may not include a representative amount of the features of the data, and there may not be any examples within the sample of all the various combinations of the variables. This was discussed earlier, and Figure 3.4 is included again here as Figure 5.4 to demonstrate stratified random sampling.

5.3.3   Sample Quality and Stability

In addition to giving insight into the dataset, measures of magnitude and dispersion can be used to check on the quality of the random sample used for detailed analysis. This can be done by calculating the mean and standard deviation of the random sample and comparing these values with those of the full dataset, also called the population. The comparison and cross

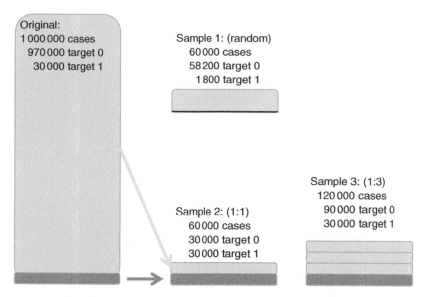

FIGURE 5.4   Stratified sampling.

tabulation tests described next can be used to assess the quality of the sample with respect to how well it represents the population.

The tests should be carried out using just some of the variables, choosing variables that are well filled with few real missing values, that is, any kind of variable for which everyone has a meaningful value, such as lifetime revenue and time since last purchase. The test variables should be relevant and important enough, as well as being representative and key to the dataset, so that if they are comparable, you can be reasonably confident that the samples are comparable.

If the sample and population are similar, then that is a reasonable indication that the quality of the sample is good. If not it is sensible to take another sample instead. If multiple samples have good quality it implies stability in the consequent models.

5.4   BASIC STATISTICS FOR PRE-ANALYTICS

5.4.1   Frequencies

Nominal, ordinal, categorical and classification variables can be investigated by looking at the frequencies of each value in a series of linked histograms. Any part of one histogram can be highlighted, and the corresponding data is

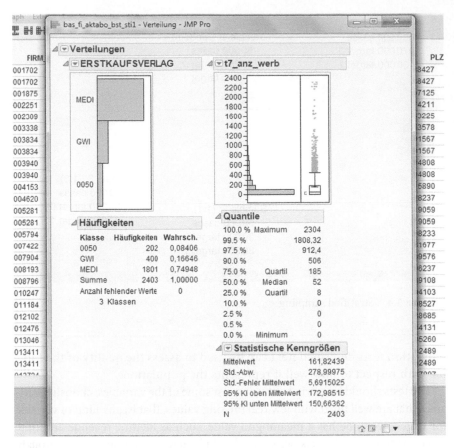

FIGURE 5.5   Example of linked histograms.

highlighted in all of the other histograms. This helps to show relationships in the data (see Figure 5.5).

Besides looking at the observed frequencies, it is often useful to look at the proportional or relative frequencies. Cumulative frequencies are also of interest to address particular questions, such as how the data is distributed in the different categories (see Figure 5.6).

Cumulative frequencies are equally applicable to both ordinal and nominal variables, for example, we could say that a person likes to ride by train as well as using a car to reach a holiday destination. Stating that the cumulative frequency of these two modes of transport is 60% has an important meaning if it is contrasted against those using a plane and a car as it may reveal a hidden pattern relating to beneficial cost savings offered or savings in travel time or reduced effort required by the customer. Note that domain

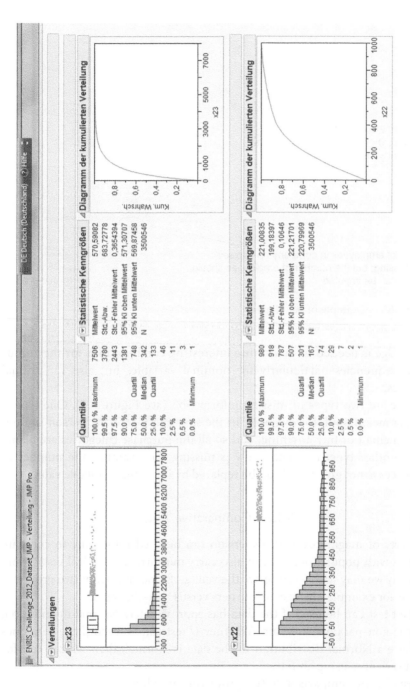

FIGURE 5.6 Example of cumulative view.

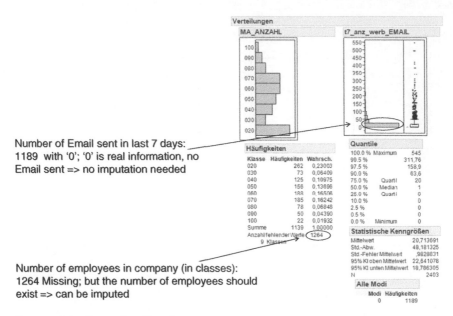

FIGURE 5.7   Example of bar charts.

knowledge is needed to determine interesting combinations for the cumulative frequencies particularly for nominal variables but also for ordinal variables.

There are two types of missing information (see Figure 5.7). One type is actually meaningful, for example, in the situation where the person did not send an email or buy anything, and so all the purchase categories are blank. For the other type, the data really is missing, for example, the number of employees is not given, and can be replaced by imputing a suitable value.

5.4.2 Comparative Tests

Measures of magnitude and dispersion can be used informally to compare samples with population. We can also carry out formal statistical tests. More generally we may wish to compare the values of a variable from two groups of people, for example, the age of churners versus non-churners.

The t test can be used if the data has approximately Normal distribution, and the non-parametric Mann–Whitney U test can be used if the data does not have a Normal distribution. If the data is Normal, then we can test for equal variances using an F test.

Tests for the comparison of two groups of variables:

t test:

> Area of application in data mining: comparison of two independent random samples concerning their average values. The values of both random samples must be approximately Normally distributed.

Mann–Whitney *U* test:

> Area of application in data mining: comparison of two independent random samples based on their central tendency. The values of the random variables can be distributed arbitrarily or have only ordinal levels of measurement.

F test:

> Area of application in data mining: comparison of the dispersion of two independent random samples. The values of both random samples must be approximately Normally distributed.

One can calculate these values in SAS with the help of PROC Univariate or PROC Means. An alternative is to let Enterprise Miner indicate this information in the data node. The different tests are also implemented in SAS.

Tests for the comparison of three or more groups of variables:

Analysis of Variance (ANOVA):

> Area of application in data mining: comparison of three or more independent random samples concerning their average values. The values of all random samples must be approximately Normally distributed.

Kruskal–Wallis test:

> Area of application in data mining: comparison of three or more in dependent random samples based on their central tendency. The values of the random variables can be distributed arbitrarily or have only ordinal levels of measurement.

5.4.3 Cross Tabulation and Contingency Tables

For variables with nominal and ordinal scaling, cross tabulation can be used to analyse the relationship between the target variable and input variables. The variables are entered into a contingency table, and a chi-square test of independence between the variables can be carried out. If the test is not significant, it indicates that the variables are independent which implies that the variable does not influence the target. If the variable and target are not independent, then this implies that the variable can be useful in predicting different levels of the target variable; in other words, the variable may be used to discriminate between different levels of the target. The contingency table may show up non-linear relationships where some levels of the input variable lead to higher levels of the target variable. This is very useful and adds more than a dependence analysis. Contingency table analysis can be extended to interval (continuous) variables as well if they are first categorised.

Tests for the assessment of the relationship between the variables:

Chi-square test of independence:
Area of application: examination of the relationship between two variables. The basic idea behind the test is whether the observed frequency of observations within the contingency table differs significantly from the frequencies to be expected (on the basis of the marginal distributions). The test only has a sensible result if more than 80% of the cells in the contingency table have expected frequency of at least 5. If necessary, categories must be combined in one or both variables to avoid cells with small frequencies.

Fisher's Exact test:
Area of application: used instead of a chi-square test if a 2×2 contingency table has very low expected frequencies ($< = 5$).

In addition to these tests, there are many other ways in which a relationship can be judged. Each test has specific assumptions, requirements and applications. Most software gives a range of different test results. Usually, all or most of the tests lead to the same conclusions (but not always, especially if it is difficult to decide what type of data you are dealing with). Details of the different tests can be found in software help files and in the Bibliography.

5.4.4 Correlations

Correlations are another possibility to determine the relationship between two variables when the variables have at least ordinal levels of measurement. For variables with nominal scale, no correlations can be calculated, and the contingency table analysis is appropriate. The parametric Pearson correlation coefficient is suitable when the data is approximately Normal or at least not too skewed, and the non-parametric Spearman rank order correlation coefficient can be used otherwise. The non-parametric test works with the ranks of the data rather than with the data itself.

The correlation is usually indicated with the letter r, which can take values between -1 and 1. If the value lies near to 1 or -1, one can assume that a strong linear relationship (positive or negative) exists. If the value lies close to 0, it can be assumed that no linear correlation exists. Correlation analysis should always be accompanied by a scatterplot of the two variables as they may be related in a non-linear way, in which case the correlation coefficient will be near to zero even though a relationship exists. An example of a non-linear relationship is age and time spent shopping because the time may be larger for teenagers and older (retired) people but less for those busy working in middle age.

Correlation adds more information than whether or not the variables are independent as it shows both the strength and the direction of the relationship.

5.4.5   Association Measures for Nominal Variables

The contingency coefficient is used in place of the correlation coefficient if both variables are nominal. The chi-square test forms the basis of the contingency coefficient. The contingency coefficient can only take values between $0 < C < 1$, the greater the relationship, the greater the value. Contingency coefficients vary with sample size and the size of the contingency table, so to compare different contingency coefficients (on the basis of different-sized samples and contingency tables), one must standardise. In contrast to the correlation coefficient, the corrected contingency coefficient does not indicate the direction of the correlation but only its strength.

Cramer's V (also called Cramer's phi) coefficient is a special form of the contingency coefficient. Its advantage lies in the fact that all values can be reached between 0 and 1, and it is not affected by sample size. Neither the contingency coefficient nor Cramer's V gives the direction of the relationship.

An overview of some correlation procedures is given in Figure 5.8.

There are other alternatives in addition to those in Figure 5.8. Each test procedure has specific assumptions, requirements and applications. Most

Scale Level	Correlation Procedures
Both variables continuous and Normally distributed	Pearson correlation coefficient (also called product moment correlation)
One or both variables ordinal-scaled or continuous but not Normally distributed	Spearman rank order correlation coefficient Kendall's Tau rank correlation (is seldom used, only with big outliers)
One binary variable (dichotomous) and one continuous and Normally distributed	Point-Biserial correlation equivalent to t-test or Mann–Whitney U test
Both variables nominal	Contingency coefficient Cramer's V

FIGURE 5.8   Overview of correlation procedures.

software gives a range of different test results. Usually, all or most of the tests lead to the same conclusions. Details of the different tests can be found in software help files and in the Bibliography.

A limitation of the comparison tests described earlier and cross tabulation is that only 'two-way relationships' can be checked; the interactions between several input variables and the target variable cannot be ascertained in this way. One possibility is to find partial correlations. Another is to find Principal Components (PCs) of several input variables and analyse the relationship between each PC and the target variable.

5.4.6 Examples of Output from Comparative and Cross Tabulation Tests

The following examples use data from several projects dealing with marketing and predictive modelling. The variable names are typically complicated as they are quite often one of 2000 or more variables in the dataset. It is sensible to include as much detail in the variable name as is needed to explain it and to avoid continually referring to look-up tables. In each example, the input variable starts with X which shows that the variable is classified rather than being the original data item.

In Figure 5.9, the input variable named xDO_GROS is the monetary value spent in a special segment of women's clothing classified into a quantile variable with six classes or sextiles as previously described. A value of zero for xDO_GROS means that the person did not buy from the segment of interest in the base period; a value of five means that they did buy. Notice that values 1–4 do not appear because buying is rare in this dataset, and the quantile borders for the original xDO_GROS values are zero for the first five quantiles. This can be seen from the fact that only 421 of the sample of 8000 (5%) people actually spent any money, and the last quantile includes the top 1/6 (16%). So it can be seen that classifying

xDO_GROS	Target = 0	Target = 1	Row totals
0	3967	3612	7579
5	33	388	421
Column totals	4000	4000	8000

FIGURE 5.9 Example of 2 × 2 contingency table.

into quantiles transforms a continuous variable (money spent) into a categorical variable.

The buying took place during the base period. The target variable is the outcome of whether a person bought (or not) after a spring fashion mailshot in the target period.

Statistics for 2 × 2 contingency table

Statistic	DF	Value Probability
Chi-square	1	315.9750 < 0.0001
Likelihood ratio chi-square	1	368.8770 < 0.0001
Continuity adj. chi-square	1	314.1973 < 0.0001
Mantel–Haenszel chi-square	1	315.9355 < 0.0001
Phi coefficient		0.1987
Contingency coefficient		0.1949
Cramer's V		0.1987
Fisher's Exact Test		
Cell (1,1) Frequency (F)		3967
Left-sided Pr < = F		1.0000
Right-sided Pr > = F		6.378E−82
Table Probability (P)		7.376E−81
Two-sided Pr < = P		1.276E−81
Sample size = 8000		

In the analysis test results, the probability is consistently less than 0.0001 which is strong evidence of a relationship between the target and xDO_GROS, implying that xDO_GROS could potentially be used to predict the target. Looking at the contingency table, it can be seen that people with xDO_GROS = 5 are more likely to be in target group 1 than those people with xDO_GROS = 0.

Note that the sample size of 8000 is very large, so even slight relationships will give a statistically significant result and have a small test probability value, $P < 0.0001$.

In Figure 5.10, the input variable, Xm12_x4_ord_prod, is the number of orders placed in the preceding 12 months from the end of the base period, classified into a quantile variable with six classes or sextiles as previously described. Note that all quantiles are now represented because numbers of

Xm12_x4_ord_prod	Target = 0	Target = 1	Row totals
0	128	123	251
1	118	130	248
2	68	103	171
3	79	82	161
4	100	78	178
5	107	84	191
Column totals	600	600	1200

FIGURE 5.10 Example of a 6 × 2 contingency table.

orders are more likely to be non-zero in the dataset. The variable name, Xm12_x4_ord_prod, shows that the variable is classified rather than the original continuous data, m12 stands for 12 months back, x4 means that four sale periods are included and ord_prod stands for number of production orders.

The orders took place during the base period. The target variable is the outcome of whether a person bought (or not) after a spring white goods mailshot in the target period.

Statistics for a 6 × 2 contingency table

Statistic	DF	Value	Probability
Chi-square	5	13.3886	0.0200
Likelihood ratio chi-square	5	13.4534	0.0195
Mantel–Haenszel chi-square	1	3.3444	0.0674
Phi coefficient		0.1056	
Contingency coefficient		0.1050	
Cramer's V		0.1056	
Sample size = 1200			

In the analysis test results, the probability is consistently less than 0.05 which is evidence of a relationship between the target and Xm12_x4_ord_prod, implying that Xm12_x4_ord_prod could potentially be used to predict the target. Looking at the contingency table, it can be seen that there is no obvious linear pattern. It is important to note that the statistical significance of a relationship is shown but not its direction.

Xt7_x3_view_tt	Target = 0	Target = 1	Row totals
0	4000	3997	7997
5	0	3	3
Column totals	4000	4000	8000

FIGURE 5.11   Example of a 2 × 2 low frequency table.

In Figure 5.11, the input variable, xt7_x3_view_tt, is whether someone visited (viewed) the website or not, classified into a quantile variable. Note that the original variable was a binary variable, and it has been reclassified into a quantile variable with only two categories (0 and 5) for consistency with other classified variables. The viewing took place during the last 7 days of the base period. The target variable is the reaction to a specific email in the target period. It can be seen from the contingency table that the response rate is very low.

Statistics for 2 × 2 low frequency table

Statistic	DF	Value	Probability
Chi-square	1	3.0011	0.0832
Likelihood ratio chi-square	1	4.1600	0.0414
Continuity adj. chi-square	1	1.3338	0.2481
Mantel–Haenszel chi-square	1	3.0008	0.0832
Phi coefficient		0.0194	
Contingency coefficient		0.0194	
Cramer's V		0.0194	

Warning: 50% of the cells have expected counts less than 5. Chi-square may not be a valid test.

Fisher's Exact Test

Cell (1,1) Frequency (F)	4000
Left-sided $\Pr <= F$	1.0000
Right-sided $\Pr >= F$	0.1250
Table Probability (P)	0.1250
Two-sided $\Pr <= P$	0.2499
Sample size = 8000	

In the analysis test results, the probability is consistently high, indicating no evidence of a relationship between the target and xt7_x3_view_tt, implying that xt7_x3_view_tt could not potentially be used to predict the target.

Looking at the results, note that the expected frequency in some of the cells is low and chi-square is not valid. Most software will tell you this in the output. The non-significance could be a misleading conclusion. For this variable, we cannot make any sensible decision at this point. This non-significant result has arisen despite the fact that the sample size of 8000 is very large.

5.5 FEATURE SELECTION/REDUCTION OF VARIABLES

As we have seen, there are typically many variables, and we may also have added to them by including new seasonally adjusted variables or variables for daily changes. An important step is to focus on the key variables by selecting out the most influential.

Pre-analytics will have already highlighted redundant variables and also likely candidates for removal because of sparseness of different values. The feature selection methods described next are also helpful to describe the data, and in some cases, variables will not be immediately discarded but will be noted as candidates for likely removal when a sample is further analysed.

However, here, we encounter a major difference between traditional statistics and practical data mining. On the one hand, it is good to have a par-simonious model with just a few significant factors, but on the other hand, this could cause problems in deployment. For example, consider the situation where the aim of the data mining is to pick out the 10% best potential customers; if the variables we have included in the model as predictors have values shared by large groups of customers, then the expected target values will be the same for large 'clumps' of customers. In this case, we will not be able to differentiate between customers and may end up being unable to pick out the top 10%. A typical example is a decision tree which shows age as an important feature, but customers are only categorised according to whether they are 'old' or 'young'; here, we still have the problem of identifying the best subset from within the large age group for our marketing activity.

5.5.1 Feature Reduction Using Domain Knowledge

An effective way to reduce the number of variables is to consider what we know about the dataset and check if any of the variables are likely to be constant throughout the data. We can use a histogram or bar chart for this.

5.5.2 Feature Selection Using Chi-Square

Another quick check for important features or variables is to carry out the chi-square test on all variables individually in successive two-way contingency tables with the target variable. The variables with the highest chi-square test value will be selected. The cut-off depends on whether you have a lot of good data and also on the degrees of freedom of the chi-square. You could use the probability border $P < 0.001$ for good data or for poor data $P = 0.05$. If you have one degree of freedom, this corresponds to having a chi-square value greater than 10; otherwise, a lower cut-off of about four could be used.

5.5.3 Principal Components Analysis and Factor Analysis

The idea behind Principal Components Analysis (PCA) is to identify the sources of variation in the data and to reduce a set of variables to a smaller number of components. It is most effective when there is strong correlation between the variables within the data. The most tangible example is where there are several variables related to general expenditure, and PCA will pick out a component which is a weighted mean of all the expenditure variables and describes the wealth of the person. Having created a component which describes overall wealth, subsequent components will pick out other aspects of the varied expenditure between people, for example, a contrast between those who spend money on sport and leisure and those who spend money on consumer items. Another component may identify single versus married people. In other words, the component has higher weight for marital status than other variables in this component. Note that binary variables often end up with a component to themselves because they are good discriminators.

Using PCs in place of variables has the advantage of reducing the number of variables but introduces a further level of complexity in that the PCs have to be interpreted. Also, it requires that the same PC analysis is appropriate in the deployment phase. Good descriptions of PCA are included in many statistics textbooks, including those in the Bibliography. PCA is also available in most statistical software packages.

Factor analysis is a method related to PCA. The interpretation is slightly different, however, as factor analysis aims to identify hidden factors within the data, whereas PCA is a more mechanical analytical process that will extract linear combinations of variables that explain the most variation in the data. Data mining usually focuses on PCA.

5.5.4 Canonical Correlation, PLS and SEM

These techniques are useful when there is more than one target variable. Both canonical correlation and Partial Least Squares (PLS) aim to identify the linear combination of variables that explains most of the variation in a linear combination of the target variables. Structural Equation Modelling (SEM) allows more complicated relationships between variables and targets with structure between variables. Bayesian network analysis also seeks to identify relationships and patterns in complex data and illustrates the links between variables showing the strength and direction of their correlations. There are many other exciting new methods, such as those involving structures like copulas, vines and non-parametric Bayesian belief networks that also comb through complex data to highlight important links between targets and input variables. These methods enable complex relationships to be identified and allow data from different sources to be compared, calibrated and integrated.

5.5.5 Decision Trees

Decision trees are a very practical way of sifting through data and finding relationships and patterns. The method is useful if the other methods previously mentioned are not possible. Decision trees are important as an analytical modelling procedure but can also be used to identify likely variables for inclusion in the modelling procedure. Decision trees will be used in many of the recipes that follow.

5.5.6 Random Forests

Decision tree analysis will always produce a result even if there are only weak relationships between the target and input variables. If it is likely that the relationships are weak and also if there is not much data available, then it can be helpful to construct a random forest. This is done by selecting multiple random samples and finding a decision tree model for each sample. The multiple random samples can be disjoint (using separate data) or be resamples (bootstrap samples) of the same data.

In Figure 5.12, the target is 'bestand_5', and the columns show results for models in each of the 15 decision trees (Baums). Notice that tree number 3 is ranked (Rang) number 1 and is the best-fitting tree.

Part of one of the trees is illustrated in Figure 5.13.

In most software, the relative importance of each variable is shown, and choosing the variables that occur most often in the forest of trees gives confidence in deciding which variables are most likely to have a real influence on the target.

Bootstrap Forest für bestand_5

Gesamtstatistiken

Einzelbäume RMSE

	r^2	RMSE	N
Training	0,379	0,3531684	945
Validierung	0,261	0,3882272	252

Kumulierte Validierung

Zusammenfassungen je Baum

Baum	Teilungen	Rang	OOB-Verlust	OOB-Verlust/N	r^2	IB-SSE	IB-SSE/N	OOB N	OOB-SSE	OOB-SSE/N
1	261	6	10,529122	0,5264561	0,5368	86,375794	0,0228507	20	3,5920546	0,1796027
2	249	7	11,183017	0,5591508	0,5051	92,281349	0,0244131	20	2,6072351	0,1303618
3	249	1	10,485898	0,3883666	0,5848	75,909993	0,020082	27	8,477019	0,3139637
4	263	10	10,283051	0,5712806	0,5388	85,865873	0,0227158	18	3,3104214	0,1839123
5	263	11	9,4449846	0,5903115	0,5091	91,263492	0,0241438	16	3,0503109	0,1906444
6	257	8	8,4176715	0,5611781	0,5250	88,759921	0,0234815	15	2,2623153	0,150821
7	257	5	5,7900368	0,526367	0,5234	89,203175	0,0235987	11	1,8217161	0,1656106
8	247	15	7,6057905	0,6914355	0,5682	81,393651	0,0215327	11	0,4600557	0,0418232
9	259	12	10,052294	0,5913114	0,5147	90,822222	0,024027	17	2,7932213	0,1643071
10	241	13	9,2837705	0,618918	0,5743	79,557937	0,0210471	15	3,7323429	0,2488229
11	259	3	8,5927126	0,452248	0,4909	94,290873	0,0249447	19	2,4565985	0,1292947
12	259	9	11,265818	0,5632909	0,5476	83,963095	0,0222125	20	6,0413044	0,3020652
13	263	4	9,390252	0,5216807	0,4954	94,181349	0,0249157	18	3,3178456	0,1843248
14	251	14	12,64303	0,6654640	0,5114	91,369040	0,0241717	19	1,9420249	0,1022539
15	261	2	9,6357144	0,4189441	0,5447	83,998016	0,0222217	23	5,0170463	0,2181324

Baumansichten

Baum1

Baum2

Baum3

Baum4

Baum5

FIGURE 5.12 Example of Bootstrap Forest I.

5.6 Time Series Analysis

Sometimes, the data we are dealing with has a time element, for example, data which records a person's purchasing activity over a number of years. Much can be learnt from such data by plotting it in time order and observing the various trends, cycles and shifts in value. Sometimes, the time gap between data items is constant, for example, monthly expenditure, and sometimes, it is variable, such as when there are intermittent purchases of a product. Generally, time series analysis requires a good series of consistent data, often with equal time gaps between the observations. Nevertheless, if the data is suitable, then time series analysis can reveal many important messages from the data.

Time series readings are likely to be autocorrelated because subsequent values are related to preceding values. For example, the weather on day 1 is highly correlated with the weather the previous day. Time series data can also show seasonal trends, for example, sales increasing in summer and decreasing in

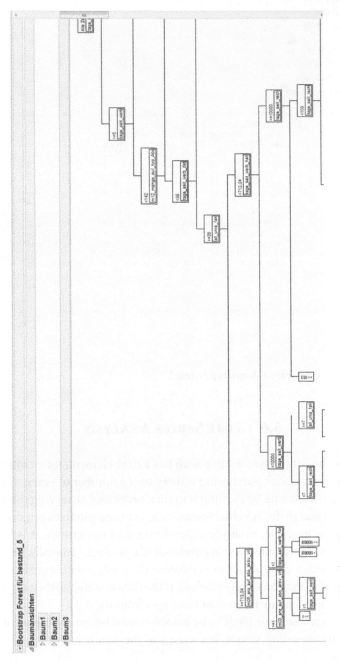

FIGURE 5.13 Example of Bootstrap Forest II.

winter. There may also be trends in that values are steadily increasing or decreasing.

One way to model time series data is to estimate the effects of trends, seasonal changes and any other cycles so that future values can be predicted. Another way is to use Autoregressive Integrated Moving Average (ARIMA) models that allow for autocorrelation. ARIMA models can also incorporate explanatory variables.

The time series models can be used to predict future customer values based on their value in previous years.

6

Methods

A Practical Guide to Data Mining for Business and Industry, First Edition.
Andrea Ahlemeyer-Stubbe and Shirley Coleman.
© 2014 John Wiley & Sons, Ltd. Published 2014 by John Wiley & Sons, Ltd.
Companion website: www.wiley.com/go/data_mining

6.1 METHODS OVERVIEW

Prediction models are generally developed using the so-called supervised learning methods. This means that models are developed using old or known inventory information consisting of input (explanatory) variables and an outcome (result or target) variable, and the development is based on a training sample of data. The 'old inventory' information can arise in the following ways:

- Responses to a previous, comparable completed campaign, for example, an autumn mailshot from the previous year.
- A buyer's model mapping the customer's reaction in a certain period or with a special seasonal event, for example, purchase in week 15 from the previous year or over Easter.
- The customers who have shown no special reaction in the base period, for example, all customers who were inactive in the first half-year of the previous year or customers who have never shopped for a particular item up to now.
- The customers who have taken part in a trial; even though these 'old inventory' files are limited in spread, they may still be capable of informing future analysis.

Unsupervised learning is an alternative to supervised learning and takes place without a target variable. Cluster analysis is an example of unsupervised learning and is well established. It is highly versatile and suitable to use with old inventory. However, because it will work with any sort of data, it is important to have some awareness of the business situation so that only sensible information is presented to the method. This is especially important because the computing time increases enormously when there are more variables. Cluster analysis is based on the idea that members of the same cluster are especially similar to themselves, but very different to members of other clusters.

A typical marketing application for cluster analysis is to derive target groups and customers' typologies from data. Here, the appeal of the application lies in the fact that one hopes to find groupings without having any prior knowledge. In department stores, for example, this technology is used to recognise typical customer groupings within the customers of a branch store to learn more about existing target groups and their profiles.

A central part of data mining is the different forecast procedures. The aim is to develop a model on the basis of old data to predict the behaviour (particularly buying behaviour) to be expected with a future application. Typical procedures are described in the succeeding text. The first sections are concerned with supervised learning prediction procedures involving multivariate analysis (classical statistical) methods followed by other data mining methods. The next sections are concerned with unsupervised learning including clustering and network methods.

Statistical procedures usually require that certain conditions are fulfilled in the data, for example, there are often assumptions about Normality, linearity or constant variance. In contrast, however, in data mining, it is expected that these conditions are not met. Data mining procedures are usually carried out with very large amounts of data (random sample sizes of 30 000 are quite common), and the non-observance of the conditions does not matter too much.

6.2 SUPERVISED LEARNING

6.2.1 Introduction and Process Steps

There are many alternative methods of analysing a dataset; they vary in the underlying concepts and in the time taken for the analysis. Different methods are appropriate in different circumstances, and in addition, people have their own preferences.

In principle, the approach to supervised learning follows the data mining process steps discussed in Chapter 2. These steps are revisited here with specific considerations for supervised learning. The process steps are:

1. Business task – Clarification of the business question behind the problem
2. Data – Provision and processing of the required data
3. Modelling – Analysis of the data
4. Evaluation and validation during the analysis stage
5. Application of data mining results and learning from the experience

Each of these process steps is described in further detail in the succeeding text.

6.2.2 Business Task

The business task step includes clarification of the problem definition, specification of aims, deciding on the planned application and the application period. A written or oral briefing by the client (industry, marketing department, etc.) is as always very important. Special attention needs to be paid to why the client really wants an analysis. It may be that it is a matter of predicting outcomes such as buying, churning or reacting in some way, or it may be that the client really wants to understand how characteristics of the customer base are related to their behaviour.

An important feature regarding clarifications and applications is to what extent the results of an analysis can be transferred and be generalised for use in practice.

6.2.3   Provision and Processing of the Required Data

To provide the required data for analysis, the following steps must be run through:

- Fix the analysis period, based on the application aim and taking into account the briefing information.

Because we are dealing here with a method of supervised learning, the temporal considerations of the variables is extremely important. Any predictive model that is generated on data from a base period needs to give results that are applicable for the deployment period. This issue was discussed at some length in Chapter 2.4.1.

- Fix the target variable.

The target variable can be continuous or categorical, and different supervised methods are discussed in the succeeding text to deal with each situation.

- Fix the population of customers to be analysed.

This is indicated by the briefing on one hand and on the other hand by the customers' attributes. For example, it may be that the client only wants to consider customers who deal with a particular branch of the company. If the analysis is going to be carried out at branch level, it must be determined whether the customers to be analysed are all customers of that particular branch.

- Generate a working random sample.

Usually, there are more than enough customers available with whom to carry out the analysis. Because of this, it is feasible and sensible to work with random samples. The supervised model is learnt or built using a learning sample of data. A sample size of approximately 30 000–100 000 has been found to be favourable in practice. A testing sample and one or more further validation samples are also selected.

- Generate input (explanatory) variables.

The input (explanatory) variables can be generated in the usual manner: investigating the data generating individual values from the facts; creating Normalised or standardised variables; using readily available variables;

creating new variables from skilful combinations of existing variables. Remember that all variables should be generated only for the base period.

According to the situation, some of the variables may need to be standardised. This is necessary particularly when the values of the variables have very different meanings, for example, consider 'amount' and 'turnover'; a value of 100 is completely different as an amount than as a turnover. It is necessary to carefully check the data levels and the additional information perhaps available in the data, and the non-information. There needs to be special consideration when the subject is 'unknown'. For example, a case is not male just because the female code has been entered as '0' rather than '1'. Such unknowns can occur where background information is entered automatically.

It is important to decide which variables must be classified and where it is important to complete data and code or recode into dummy variables. According to the procedure to be applied and depending on the desired statements, it is also important to check whether continuous variables would be better classified or not.

There are two ways to approach the classification:

In Enterprise Miner, in the area Modify, every variable has to be worked on individually by hand. The advantage of such a very individual classification has the disadvantage of a very high lead time.

With a standard SAS programme, everything can be converted into class variables by dividing the data into sextiles whose class borders are fixed by the programme.

In addition, attention needs to be given particularly to outliers. These are to be removed if necessary or eliminated. A sensible option to fix how missing values will be handled in the procedure is to state that there are 'no missings' in the records.

6.2.4   Analysis of the Data

The first step is to carry out a descriptive analysis of the data. Then, on the basis of these results and the briefing, the variables are classified if necessary, and appropriate methods and variables are chosen. Whether the aim is to understand or to predict, there are a variety of methods that can be used. Within each of the methods, there are many options for customising the analysis and different ways for the results to be presented. The choices can be quite overwhelming. Three of the most commonly used methods are described next: linear and logistic regression, decision tree analysis and neural networks. The pros and cons of these methods are listed to summarise the methods and aid the choice of which method to use. All of the methods described are implemented in JMP or SAS.

The statistical methods are introduced only briefly here to allow an overview. Details of the single methods and their applications are readily available in statistics books (see Bibliography).

Most of the supervised methods include ways to select the important variables so all the input variables can be made available for building the model. Software usually provides the facility of comparing alternative models and methods of model building and presenting the results. This is very useful but should not be used in place of thinking out which methodology might be the most appropriate for the data and the business question.

Besides considering the choice of methods, we have to check if there is any default for the sample size used with the analysis. A standard random sample size may be built-in and may be highly conservative, such as 2 000; we may prefer to state our own larger sample size, for example, 30 000, which is appropriate for the particular analysis. However, recall that the larger the sample size, the longer the analysis will take.

6.2.5   Evaluation and Validation of the Results (during the Analysis)

Supervised learning produces a model, and the value of the model depends on how well it either explains or predicts the patterns in the dataset. The process of carrying out an analysis can often be immensely valuable simply because of the focus on collecting a clean, reliable set of data. Sometimes, the benefit of the model can be demonstrated in terms of expected financial savings or increased profits.

The quality of the model can be checked by applying the model to a new sample of data and comparing predicted target outcomes with observed target outcomes. The methodology for assessing the quality depends on the supervised learning method being used and is described for each of the methods in the succeeding text.

Another way to check the validity of the model is to compare the results with what is already known about the data and business structures behind it. One should always look at the results from the pure business point of view and check that they are reasonable.

6.2.6   Application of the Results

The aim of the supervised learning is to produce a model that can be used to understand or predict future outcomes. Models can be represented by formulas or by sets of rules depending on the method. The model can be applied to the full population or to a new set of data.

6.3   MULTIPLE LINEAR REGRESSION
FOR USE WHEN TARGET IS CONTINUOUS

6.3.1   Rationale of Multiple Linear Regression Modelling

Multiple linear regression is used to estimate continuous target variables such as turnover or customer expenditure. If the target variable is binary or categorical, then logistic regression is the better choice. Multiple linear regression is recommended when we are interested in the individual predicted or intermediate values rather than those rounded to Target = 0 or Target = 1.

Linear regression is one of the easiest methods to understand and to apply; it can give very accurate and precise predictions. To derive a simple linear regression model where there is only one predictor variable, data is represented as dots in a two-dimensional model, where the Y-axis represents the target and the X-axis accommodates the predictor. Provided the dots fall approximately into a straight line rather than a curve of some sort, then a linear (straight line) regression line is drawn across the plotted data in such a way that the distance between the plotted points and the line itself is minimised. The line that has the shortest total squared distance from the plotted points is used for the predictive model. A simple linear regression model might appear like the following: prediction = $a + b \times$ predictor.

The world that we live in is not as simple as a linear model. Often, data analysts will encounter complex patterns of data that prevent them from drawing a straight line perfectly. The relationship between target and predictor may be curved, or more than one predictor may be needed to explain the variation in the target. Multiple linear regression is when additional predictors are added to produce multi-dimensional hyper-planes instead of straight-line models. Multiple linear regression models use more information and thus can often produce an enhanced prediction.

Multiple linear regression is one of the most frequently used forecast procedures. A target (or dependent) variable is estimated by several independent input variables. The target is referred to as the dependent variable because its value depends upon the values of the predictor variables. The predictor variables are also referred to as explanatory because they explain the variation in the target; they are also called independent variables, not because they are independent of the target variable but because they can take any value.

In regression analysis, the target variable and the input variables should have numerical values. Nominal variables such as 'colour' must be converted to indicator variables which usually take the values 0 and 1. Each colour has an indicator variable which indicates whether the product is that colour or not. Categorical variables such as 'size' cannot be used as numbers like 1, 2, 3 and 4 unless the intervals between the numbers are constant and a single regression coefficient can be applied to them. Usually, categorical variables must be converted to separate indicator variables, one for each category.

A multiple regression model might have many dimensions. The general formula for the multiple linear regression model is

$$Y = a + b1 \times X1 + b2 \times X2 + \ldots + bn \times Xn$$

where:

Y = target variable

$b1 \ldots bn$ = the parameters/regression coefficients which are estimated in the model

$X1 \ldots Xn$ = input variables

a = constant

A regression analysis with only nominal input variables is effectively an Analysis of Variance (ANOVA). A regression with a mixture of continuous and nominal variables is an analysis of covariance or equivalently a regression analysis with continuous and indicator variables.

Regression makes use of important variables to predict the target. Depending on the problem, there will be many different input variables. In addition, it is sometimes useful to include some variables that have no proven relation to the problem but are widely reported, as these variables can give an overview of the general environment. For example, we might include the channel of placing an order so that we can report the current situation even though it bears no relationship to whether the customer responds to an advertising campaign or not.

It is necessary to learn about the distribution of the input variables and transform them if necessary. The benefit of approximately Normal input variables is that the spread of input values is assured which leads to more precise estimates and better predictions. We can standardise the input variables to give them a comparable scale or convert to quantiles. This can help to make alternative models easier to compare but quantiles should only be used if there are a lot of cases and variables because quantiles necessarily mean a loss of information.

6.3.2   Regression Coefficients

For each of the regression coefficients, a test size T (for t test) is calculated by dividing the coefficient by its respective standard error, and the corresponding p value is found. A p value less than the significance level, for example, $p < 0.05$, indicates that the regression coefficient is significantly different from zero and the input variable has an effect in the model. On the other hand, if the regression coefficient is not significantly different from zero, then the input variable has no effect in the model.

A multiple linear regression can be carried out in two ways:

- The model uses all input variables.
- The model uses only the relevant input variables selected by Stepwise, Backwards or Forwards procedures.

The second option is more common and is appropriate when there are many input variables available. According to a set procedure, the contribution of each potential input variable is checked every time another variable is added to or taken away from the model to see whether it gives additional explanation to the prediction of the target variable. The user decides on the significance level for adding input variables and for removing input variables. A significance level of 15–20% is not uncommon.

Non-stepwise regression models will enter all variables together. Presenting the variables in order of importance ensures that the algorithm gives the best chance of being entered in the model to the most important variables. The order we use includes not only statistical considerations such as the order of importance in terms of correlation but also the more business-focused considerations of presenting more economical or reliable variables first so that the most useful variables have a good chance to be included in the model.

If models of similar power are found which have alternative variables (due to correlation between the variables), then the cheaper or easier or more reliably measured variables should preferentially be included over the alternative variables. Looking at the correlation between input variables helps determine variables that can act as surrogates or proxies to each other.

6.3.3 Assessment of the Quality of the Model

Data mining models aim to pick out the features which explain variation in the target variable. The extent to which the model achieves this objective is assessed in a number of ways.

An important check on the model is to examine the discrepancies between the observed target values and those predicted by the model. These discrepancies are called residuals, and they give important information about the fit of the model and can be used to diagnose possible problem areas. Information about most of these problem areas can be obtained from a visual exploration of the residuals. After fitting a regression model, diagnostic plots of the residuals are examined to look for patterns.

Diagnostic plots include:

- A plot of residuals in case order. We want the model to fit early and later cases equally well, and if this is the case, the residuals will vary randomly when plotted in case order with no trend or cycle.
- A histogram. We want the residuals to have an approximately Normal distribution, and so, the histogram should be symmetrical with a single peak around zero.
- Scatterplots with residuals on the vertical axis and either predicted values or any of the explanatory variables on the horizontal axis. We want the model to fit equally well for large and small predicted values; also, we want the model to include all important aspects of the explanatory variables. In these cases, scatterplots will show no trend or other pattern.

If any patterns are evident in the diagnostic plots, it implies that the model can be improved, either by transforming one or more of the variables or by adding additional variables into the model.

An overall measure of fit is conveniently summarised in the $\%R^2$ value. This is the % of the variance of the target variable which is explained by the model. In statistics, this value should be reasonably large for the analysis to be meaningful, for example, values more than 60% are considered as satisfactory. However, in data mining, particularly when estimating buying behaviour, these values are practically never reached. Here, a $\%R^2$ value as low as 11% might be considered very good. Even very low $\%R^2$ can indicate a useful model as there is at least some indication of a relationship between input and target variables even though it may not be very strong.

According to experience, one should be very critical with a model for, say, buying behaviour, if the $\%R^2$ value reached is greater than 20%. The reason for this is that the noise in the data usually leads to low $\%R^2$ values, and so, if a value greater than 20% is reached, it is almost always the case that one or more of the input variables separate the target cleanly in the period, for example, if the input variable is purchase location United Kingdom or Germany and the target variable is buy or not buy, then a large $\%R^2$ could arise because the product is only sold in the United Kingdom.

Also, there may be overlap between input and target, for example, one input variable could be a flag for a certain behaviour occurring, and it could happen that the behaviour took place at the same time as the target was achieved. For example, this can happen if you have summarised data without a time stamp. For example, if a variable represents how many cars someone has bought but it is not time stamped, then it is very likely that the last car they bought is also included in the data summation. If the target variable is buy or no buy this particular new car, then the number of cars bought is likely to be highly correlated with the target producing a model with a high $\%R^2$ value. The solution is to be more careful about subtracting the car of interest from the summation (and any other cars bought after this one). To find candidate variables for this sort of problem, look at the importance ranking of each variable given by the data mining tool or the % of contribution. You should find the useful candidate within the first three variables, and it will have an enormous contribution to the model which might be suspicious and should be cross-checked with domain knowledge. There is still a chance that the model is working and is a good model, but it has to be checked. An easy method to check is to construct a contingency table for the target and the suspicious variable, and if the table has nearly 100% frequencies on the diagonal, then check the timeline very carefully. Note that if the model is being used for explanation or description, it is not really a problem and is in fact an interesting finding; however, if the model is being used for prediction, it is unlikely to work on new data.

6.3.4   Example of Linear Regression in Practice

The target variable used in linear regression must be continuous, and the input variables must be continuous or ordinal. In the JMP Pro application in Figure 6.1, the target variable Y is chosen to be the variable called x23.

In Figure 6.2, the analysis method is chosen and the explanatory variables are added.

The target variable must be checked to be at a suitable measurement level for the method chosen. In Figure 6.3, it can be seen that there are two variables that could be useful targets. The variable x23 is a continuous variable and 'target_female' is a binary variable.

As target_female is binary, it is suitable as a target for logistic regression, decision tree analysis or neural networks. As x23 is a continuous variable, it is suitable as a target for linear regression. It should be noticed that the distribution of x23 is very skewed, and it may be better to take a log transformation and use that as the target or put x23 into bins and use a logistic regression. However, for the sake of this example, we will continue with x23 as the target without transformation.

In the regression model, significant variables are identified using the stepwise approach. In Figure 6.4, only variables x6 to x9 and x1001 have a

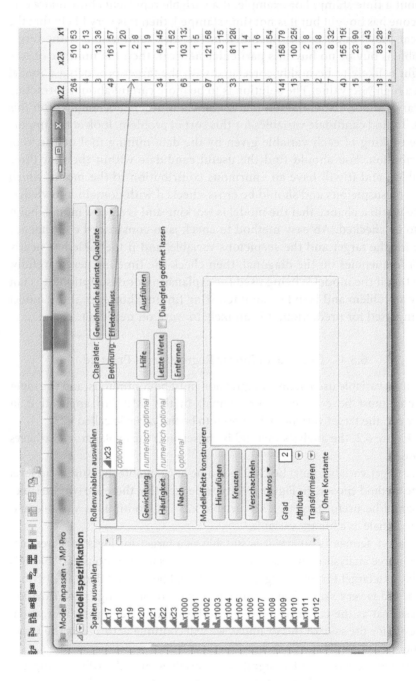

FIGURE 6.1 Linear regression model and choice of target variable.

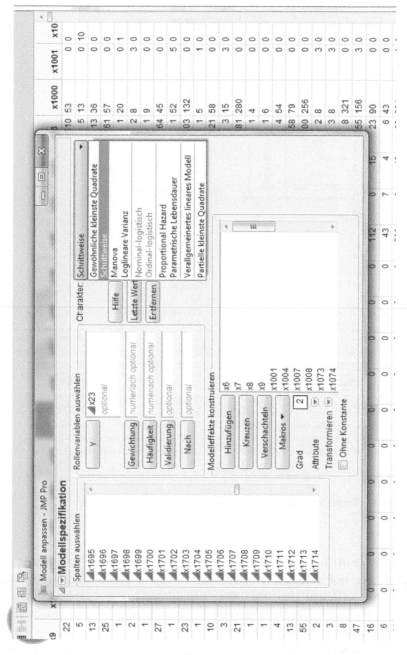

FIGURE 6.2 Choice of the analysis method and explanatory variables.

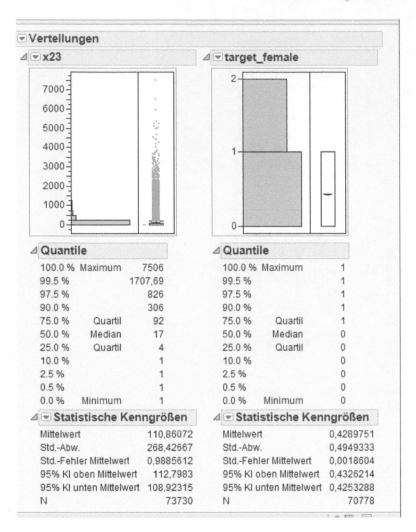

FIGURE 6.3 Target variables being checked.

significant contribution to the model and are retained. The other input variables are not selected. Notice that x9 has a much larger estimated regression coefficient than the other variables. This is a typical situation arising with real data containing noise and lots of variability but one or more strongly related explanatory variables.

The significance levels for the stepwise approach must be fixed. As given in the example earlier, it works well if the significance level to enter the model is

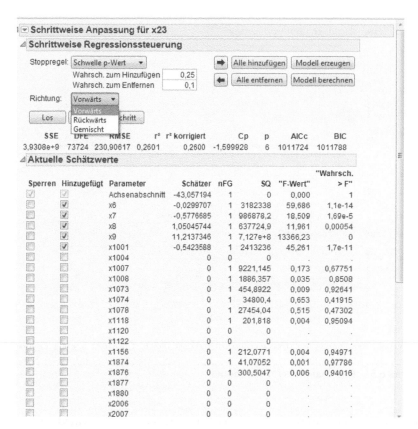

FIGURE 6.4 Reduction of the number of input variables using the stepwise approach.

less demanding than the one to stay in the model. In the example, the entry level is 0.25 (25%) and the level to stay in the model is 0.1(10%).

If it is possible to pre-sort the list of input variables by expected importance in the model, then this will improve the model. This prior knowledge can be gained by experience or acquired as a result of previously carried out analytics.

In the stepwise approach, the regression coefficients are only calculated for input variables that make a significant contribution to the model. The parameters are listed in the output in Figure 6.4.

Most data mining software offers options to optimise the modelling process, but more often, the default settings are sufficient and are definitely a good starting point.

The results of regression modelling are enhanced by graphical presentation. Different charts are used depending on the nature of the target variable. If the

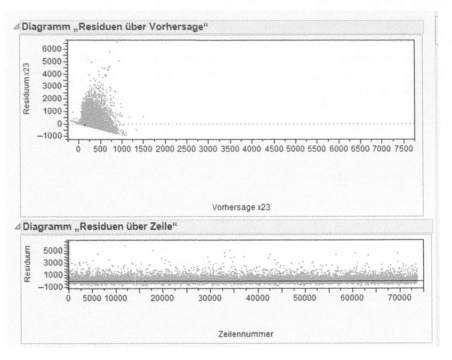

FIGURE 6.5 Residual plots for the model.

target variable is continuous, then no lift or gain charts are provided nor is there a confusion matrix.

Comparison of the real and predicted outcomes can usefully be shown in a scatterplot. The more the points lie on a 45° line, the better the model. The model is not very good if the plot shows a pattern or a random scatter.

Recall that the difference between the observed and predicted outcomes is called the residual. There is a residual value for each case. In Figure 6.5, the residuals are plotted on the vertical axis with the predicted values on the horizontal axis in the upper plot and the case number in the lower plot.

Ideally, both of the aforementioned plots should show a random scatter of points as this would demonstrate that the model fits equally well for all levels of the predicted outcome and also for all cases. The lower plot shows a reasonably random scatter, but there is a pattern to the upper plot. The model in Figure 6.5 is not particularly good.

The statistical output differs a lot from data mining tool to data mining tool.

Detailed results for a stepwise linear regression in an SAS Output are shown in Figure 6.6.

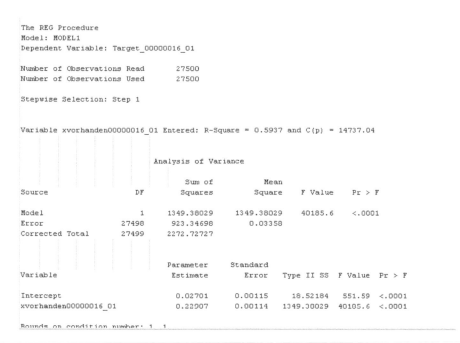

```
The REG Procedure
Model: MODEL1
Dependent Variable: Target_00000016_01

Number of Observations Read        27500
Number of Observations Used        27500

Stepwise Selection: Step 1


Variable xvorhanden00000016_01 Entered: R-Square = 0.5937 and C(p) = 14737.04


                                Analysis of Variance

                                  Sum of         Mean
Source                    DF      Squares       Square    F Value    Pr > F

Model                      1   1349.38029   1349.38029    40185.6    <.0001
Error                  27498    923.34698      0.03358
Corrected Total        27499   2272.72727


                            Parameter    Standard
Variable                    Estimate       Error   Type II SS  F Value  Pr > F

Intercept                    0.02701     0.00115     18.52184   551.59  <.0001
xvorhanden00000016_01        0.22907     0.00114   1349.30029   40185.6  <.0001

Bounds on condition number: 1   1
```

FIGURE 6.6   Detailed results for a stepwise linear regression Part 1.

After the constant is added to the model, the relevant X variables are considered step by step. The final model is shown in Figure 6.7.

It often happens that a model contains quite a lot of variables, but only the first ones have a really big influence on the model.

6.4   REGRESSION WHEN THE TARGET IS NOT CONTINUOUS

6.4.1   Logistic Regression

In the business realm, logistic regression is widely used where the target outcome is binary, for example, just YES or NO, or where the target outcome is categorical with just a few categories. It is an alternative to linear regression which is used to model continuous target variables. Rather than making a prediction of outcome corresponding to a given set of predictors, logistic regression calculates the probability of the outcome occurring. The transformation logit (p) is used in logistic regression with the letter p representing the probability of success. Logit (p) is a non-linear transformation and logistic regression is a type of non-linear regression (see Figure 6.8). The expected

All variables left in the model are significant at the 0.1000 level.

No other variable met the 0.1000 significance level for entry into the model.

Summary of Stepwise Selection

Step	Variable Entered	Variable Removed	Number Vars In	Partial R-Square	Model R-Square	C(p)	F Value	Pr > F
1	xvorhanden00000016_01		1	0.5937	0.5937	14737.0	40185.6	<.0001
2	xvorhanden00000000_03		2	0.0327	0.6264	11338.1	2408.22	<.0001
3	xvorhanden00009016_01		3	0.0144	0.6408	9847.63	1099.00	<.0001
4	xm1_anz_werb_sol		4	0.0067	0.6475	9149.85	525.12	<.0001
5	xerfassungsart00000016_02		5	0.0037	0.6512	8770.02	289.54	<.0001
6	xvorhanden00009280_01		6	0.0033	0.6545	8425.97	264.93	<.0001
7	xm3_anz_werb_Fe		7	0.0031	0.6576	8107.32	247.68	<.0001
8	xvorhanden00000099_01		8	0.0021	0.6597	7890.42	170.13	<.0001
9	xm48_anz_werb_ma1		9	0.0019	0.6616	7693.41	155.54	<.0001
10	xvorrat00000016_01		10	0.0013	0.6629	7564.62	102.60	<.0001
11	xOverall_Recency00000016_01		11	0.0068	0.6697	6857.51	567.73	<.0001
12	xretouren00000016_01		12	0.0025	0.6723	6594.93	213.46	<.0001
13	xGeschlecht		13	0.0010	0.6732	6494.79	82.65	<.0001
14	xOverall_Recency00000071_03		14	0.0007	0.6740	6421.98	60.67	<.0001
15	xvorhanden00002003_02		15	0.0007	0.6747	6352.01	58.49	<.0001
16	xvorrat00000792_01		16	0.0007	0.6753	6285.55	55.74	<.0001
17	xvorrat00009016_01		17	0.0007	0.6760	6218.46	56.37	<.0001
18	xretouren00009016_01		18	0.0008	0.6768	6136.76	68.46	<.0001

FIGURE 6.7   Detailed results for a stepwise linear regression – Final model.

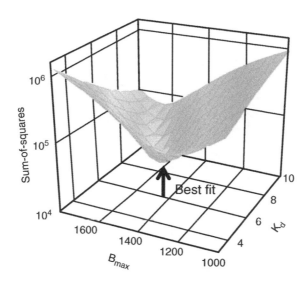

FIGURE 6.8 Non-linear regression.

values of the target variable from a logistic regression are between 0 and 1 and can be interpreted as a likelihood.

The statistics values for the assessment of the model quality are analogous to those in multiple linear regression, but instead of the test size *T*, the test size Wald is shown. The test size Wald is the statistic for the examination of the significance of the regression coefficients. The Wald statistic is the square of dividing the coefficient by its respective standard error and follows a chi-square distribution. The greater the value of the statistic, the stronger the influence of the variable is on the model.

On the plus side, it is easy to model using logistic regression with standard software. The downside of this method is that it requires a large amount of manual work if there is no Stepwise option available or feature reduction methods cannot be carried out in advance. It also makes assumptions about distribution and linearity. In comparison to other methods, regression methods are more sensitive to outliers, and performance is not always outstanding.

6.4.2   Example of Logistic Regression in Practice

Analogously to linear regression, the steps are illustrated in JMP in Figure 6.9.

To carry out a logistic regression, select 'Logistic' and 'Logit' in the model options node.

In the second step, the target variable is checked as shown in Figure 6.10.

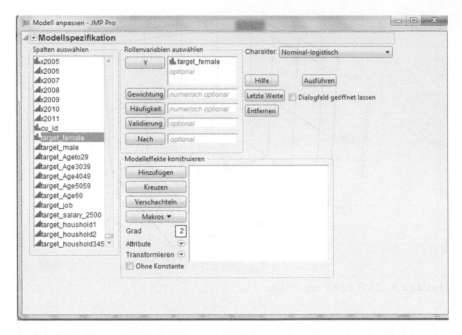

FIGURE 6.9   Non-linear regression steps Part 1.

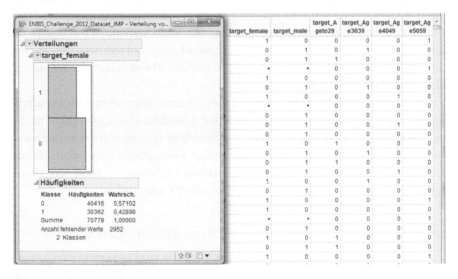

FIGURE 6.10   Non-linear regression steps Part 2.

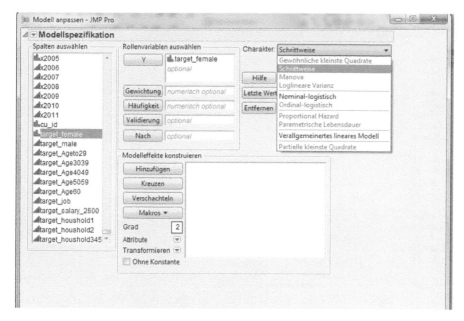

FIGURE 6.11   Non-linear regression steps Part 3.

In Figure 6.11, it is shown that one can select the procedure to be used analogously to linear regression.

The different significance levels and choice conditions can be fixed under 'Criteria' as shown in Figure 6.12. In special cases, it can make sense to put the significance level as high as 15% or to choose different levels for entering the model and staying in the model.

It is important to recheck whether all the input variables are at the right level. Sometimes, as in Figure 6.13, a variable is stored in the wrong format.

The variable x1039 should be numerical but is stored as a character. If it stays as a character, the software will either reject the variable or it will create a new dummy variable out of each value of the variable. This might have a bad effect on the model and also on the processing time.

We recommend that the pre-sets are usually retained for the model optimisation unless you are an experienced data miner or statistician. Typical output is shown in Figure 6.14.

Lift and gain charts as well as the confusion matrix are available to help with the interpretation of the results for the binary target variable analogously to the decision trees. A logit model can be compared together with decision trees or other logit models or neural networks in the assessment node.

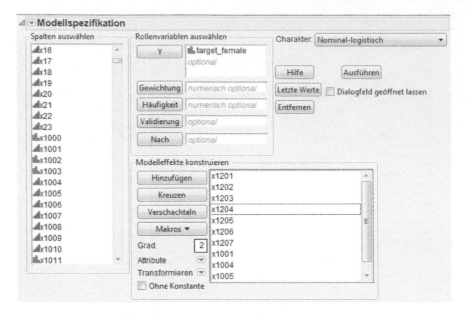

FIGURE 6.12   Non-linear regression steps Part 4.

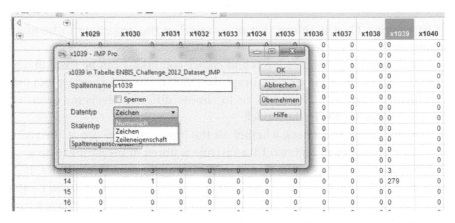

FIGURE 6.13   Non-linear regression steps Part 5.

Nominal variables are represented by indicator variables as in Figure 6.15. It should be cross-checked whether indicator variables are appropriate for the model or whether it would be better to change the variable to a different scale. Analogously to the linear regression, the parameters are shown only for the variables that are relevant for the model.

FIGURE 6.14   Non-linear regression steps Part 6 results.

Different software gives different diagnostic plots. The examples in Figure 6.16, Figure 6.17 and Figure 6.18 are from JMP. Figure 6.16 shows different ways of assessing the quality of the model. It includes analysis of residuals.

Figure 6.17 looks at the fit of the model and includes an ROC curve. Instead of the 45° line from (0,0) to (1,1), a tangent is drawn where the curve is greatest. In Figure 6.17, the curve is not very marked, suggesting that the model is not particularly good.

Figure 6.18 gives an overview with different fit statistics. The statistics are calculated for the training sample and the test (validation) sample. The table enables a comparison for each statistic to be made and shows in what ways the fit of the model differs between training and validation samples. This indicates

Term		Schätzer	Std.-Fehler	Chi²	Wahrsch.>Chi²
x1002[1-0]		0,5201773	0,0882043	34,78	<,0001*
x1002[10-1]		0,42086764	0,5118355	0,68	0,4109
x1002[100-10]	Instabil	22,0599088	109201,97	0,00	0,9998
x1002[102-100]	Instabil	-0,6870844	154330,81	0,00	1,0000
x1002[103-102]		0,69164271	154390,73	0,00	1,0000
x1002[104-103]		-0,5214009	154305,22	0,00	1,0000
x1002[106-104]		0,27841638	133530,75	0,00	1,0000
x1002[108-106]	Instabil	-23,716968	76976,439	0,00	0,9998
x1002[109-108]	Instabil	24,0093763	109294,23	0,00	0,9998
x1002[11-109]	Instabil	-22,639769	109294,23	0,00	0,9998
x1002[111-11]	Instabil	22,451201	109379,29	0,00	0,9998
x1002[112-111]	Instabil	-0,0629715	154589,31	0,00	1,0000
x1002[116-112]		-0,1642175	154627,78	0,00	1,0000
x1002[117-116]	Instabil	-0,2786344	154526,2	0,00	1,0000
x1002[118-117]	Instabil	-22,717137	109135,46	0,00	0,9998
x1002[12-118]		2,78932091	1,7862333	2,44	0,1184
x1002[128-12]	Instabil	19,8133636	77296,2	0,00	0,9998
x1002[129-128]	Instabil	-45,825671	134000,46	0,00	0,9997
x1002[13-129]	Instabil	24,8897781	109386,65	0,00	0,9998
x1002[130-13]	Instabil	-25,277176	109155,64	0,00	0,9998
x1002[135-130]	Instabil	46,3558811	133630,43	0,00	0,9997
x1002[136-135]		0,10793939	133669,59	0,00	1,0000
x1002[14-136]	Instabil	-21,996505	109203,31	0,00	0,9998
x1002[141-14]	Instabil	22,3475335	77219,757	0,00	0,9998
x1002[15-141]	Instabil	-21,781408	77219,757	0,00	0,9998
x1002[152-15]	Instabil	21,4365537	109201,98	0,00	0,9998
x1002[157-152]	Instabil	-22,836397	109201,98	0,00	0,9998
x1002[159-157]	Instabil	22,1875855	109270,03	0,00	0,9998
x1002[1598-159]	Instabil	0,02578149	154376,93	0,00	1,0000
x1002[16-1598]	Instabil	-21,132619	109166,3	0,00	0,9998
x1002[165-16]	Instabil	21,767897	109449,35	0,00	0,9998
x1002[169-165]	Instabil	-46,507995	154723,19	0,00	0,9998
x1002[17-169]	Instabil	25,6974899	109342,35	0,00	0,9998

Above table headed by:

⊿ ▼ Nominal-logistische Anpassung für target_female
⊿ Parameterschätzer

FIGURE 6.15 Non-linear regression steps Part 7.

whether the model can be generalised or not. In Figure 6.18, the results are fairly similar, implying that the model is not too bad in this sense.

As well as the statistical measurement, further evaluation and validation are needed.

6.4.3   Discriminant Analysis

Discriminant analysis determines the combination of the input variables which leads to a maximum separation between the classes of the target variable. In this respect, discriminant analysis resembles logistic regression.

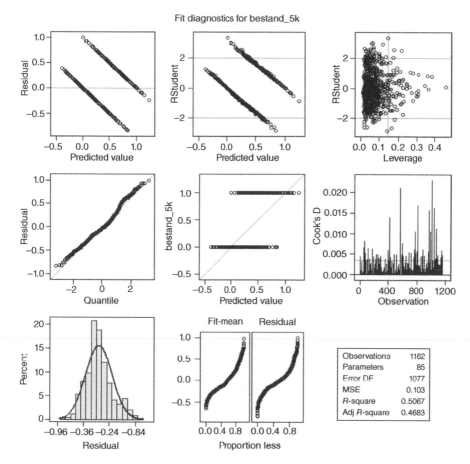

FIGURE 6.16 Non-linear regression options to measure the model quality.

In general, for the discriminant analysis to be valid, the number of cases or observations (typically customers) must be bigger than the number of input variables, and the number of input variables should be bigger than the number of classes in the target variables. Normally, these conditions are fulfilled. However, Normal distribution assumptions are part of this method, and real data often causes problems with these assumptions, making the results unreliable in practice. From our experience, regression is more robust and therefore preferable. However, if logistic regression is not available in the analytic software, then discriminant analysis provides an alternative.

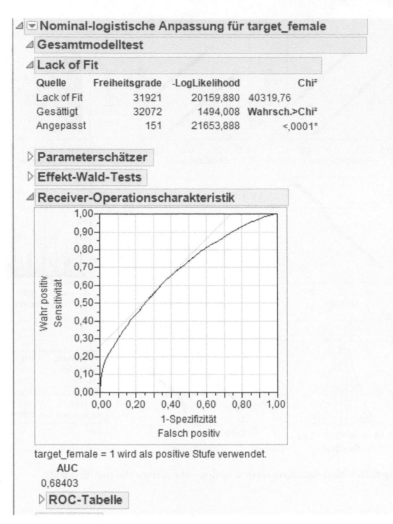

Figure 6.17   Non-linear regression steps Part 8.

6.4.4   Log-Linear Models and Poisson Regression

Log-linear models are generally used to develop prediction models with nominal-scaled and ordinal-scaled input variables and the combinations of the different variables (interactions). The target variable for these models is also nominal or ordinal. The model can include the main effects of the individual variables if they are looked at independently of each other, and also the interaction effects between input variables. The models can be saturated or unsaturated. A model is saturated if all interaction effects are included in the analysis. If only a subset of interaction effects is included, then the model is unsaturated.

Fit Statistic	Label	Training	Validation	Test
AIC	Akaike's Information Criterion	5099.5962443	.	.
ASE	Average Squared Error	0.1939015325	0.1867942489	.
AVERR	Average Error Function	0.5724460253	0.5571592332	.
DFE	Degrees of Freedom for Error	4424	.	.
DFM	Model Degrees of Freedom	11	.	.
DFT	Total Degrees of Freedom	4435	.	.
DIV	Divisor for ASE	8870	5914	.
ERR	Error Function	5077.5962443	3295.0397052	.
FPE	Final Prediction Error	0.1948657806	.	.
MAX	Maximum Absolute Error	0.9642053011	0.9750825591	.
MSE	Mean Square Error	0.1943836566	0.1867942489	.
NOBS	Sum of Frequencies	4435	2957	.
_NW_	Number of Estimate Weights	11	.	.
RASE	Root Average Sum of Squares	0.4403425173	0.4321970024	.
RFPE	Root Final Prediction Error	0.4414360437	.	.
RMSE	Root Mean Squared Error	0.4408896195	0.4321970024	.
SBC	Schwarz's Bayesian Criterion	5169.9663562	.	.
SSE	Sum of Squared Errors	1719.9065934	1104.701188	.
SUMW	Sum of Case Weights Times Freq	8870	5914	.
MISC	Misclassification Rate	0.2874859076	0.2675008455	.
PROF	Total Profit for ZIEL	3080	2082	.
APROF	Average Profit for ZIEL	0.694475761	0.7040919851	.

FIGURE 6.18   Non-linear regression steps Part 9.

Poisson regression is used when the target variable is count data such as the number of complaints a customer makes. A model is developed that uses the input variables to explain the number of complaints. Usually, if the counts are fairly wide ranging, linear regression would be a good approximation, but if there are few options for the target count variable such as when the target is the number of offers a customer responds to, then Poisson regression may be preferable. This methodology is also applied in Rasch Measurement Theory used widely in education and social science studies. See Bibliography.

6.5   DECISION TREES

6.5.1   Overview

Decision tree analysis is one of the most popular and most reliable data mining methods for developing prediction models. The basic idea behind it is 'divide and rule'.

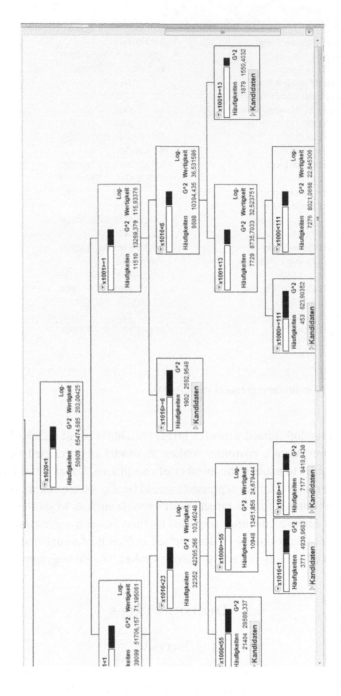

FIGURE 6.19 Decision tree.

A decision tree is a predictive model and as its name implies, it can be viewed as a tree. Each branch of the tree represents a unique classification question, and all leaves of the tree are partitions of the dataset with their classification. The division of the data must fulfil the mutually exclusive and collectively exhaustive rules. Mutually exclusive means that the partition of data into leaves must be distinct, that is, one piece of data can only appear in one of the leaves on that branch. A collectively exhaustive decision tree contains all the dataset divided accordingly into many leaves (see Figure 6.19).

Marketing and sales people are well accustomed to segmenting data into groups of customers, products or sales regions to obtain a high-level overview of large amounts of data for general use. These groups of data are then managed separately for promotional purposes or to gauge the level of acceptance of certain products. Conversely, decision tree segmentation is carried out for the specific purpose of prediction. The data is segmented according to a predicted outcome, and the data is classified in each segment, or leaf, because they comply with or have similarities with the predicted outcome and not just because they have general similarities. Therefore, the method of data division enables the intended information from the decision tree to be easily assimilated and comprehended. Due to this clarity, the decision tree also allows for more complex profit or Return On Investment (ROI) models to be added on top of the predictive model. For instance, once a large group of customers with high certainty or likelihood to purchase certain products is found, complex costing or profit and loss models could be utilised, on top of the existing decision tree predictive model, to gauge the suitability of a particular marketing instrument to ensure a handsome ROI.

Decision trees are relatively easy to build, since they were originally developed to mimic the way human beings think and solve problems. Therefore, they are particularly adapted to handling large amounts of data with minimal assumptions on data partitions. The final predictions are easy to understand and subsequently are easy to implement. The first step in defining the tree is to agree on the suitable question for the root and subsequently for all of the branches. For example, the root question or outcome of interest may be whether a customer churns or not. The whole set of data is first partitioned at the root, and recursively, the division of the data is done by moving downwards, that is, growing the tree, according to the selected partitioning attribute. At all stages, the tree partitions are aimed at answering the root question, for example, which customers are more likely to churn.

At the root of the tree, the partitioning is easier to do, because the suitable patterns for splitting are readily identifiable, for example, the first branch may have males in one leaf and females in the other leaf. As the process gets to the smaller leaves, that is, when the tree starts growing, the pattern for splitting gets less distinct, and the selection algorithm tries to over-fit the data into their respective leaves. For example, one of the smaller branches may have people who last bought in summer on one leaf and those who last bought at a different time on the other leaf. The result of this sort of over-fitting is an unstable tree that will not produce a good prediction. As a solution to weed out these outlying divisions and identified 'noises', a process called pruning is conducted to eliminate the unstable leaves by merging them. Most of the algorithms for pruning rely on adjusted mathematical error rates. The larger the tree grows, the weaker the leaves get. A threshold is applied to these leaves, and the weaker links are merged or pruned.

In summary, a decision tree is used for classification and prediction, and it combines both data exploration and modelling. A decision tree model consists of a set of rules to divide heterogeneous data into smaller, more homogeneous groups of data with respect to a particular target variable. The rules can be expressed in English, or any other language, and presented in a way for easy comprehension. For example, a simple rule could be that males who last bought in summer are most likely to churn.

The rules could also be expressed as strings of SQL statements to retrieve data of a particular category from databases. For example, the rule could be as follows: select males who last bought in summer. If the tree is used to extract rules, it is better if the tree is kept short or if a more complex tree is collapsed into smaller sets of leaves to produce readable and actionable rules. As an illustration, examine the following examples which have been extracted from a decision tree with the target variable being happiness:

The first branch is as follows: watch the soccer game or not. Thereafter, happiness increases via alternatives including friends, home team result, beer, whisky, pizza. The resulting rules for happiness are:

- Watch the soccer game and home team wins and go out with friends and then have a beer.
- Watch the soccer game and home team wins and sit at home and then have a pizza.
- Watch the soccer game and home team loses and go out with friends and then have a beer.
- Watch the soccer game and home team loses and sit at home and then have a whisky.

In the earlier example, *watch the soccer game?* is the root of the tree as it repeats itself in all of the rules. Next, if the answer to the root question is *yes*, then the subsequent question to split the data will be *home team wins?* As the flow goes, the remaining questions are applied to further split the data into homogeneous groups with beer, pizza and whisky being the final leaves. Moreover, the rules that end with *beer* could be merged to produce a new and simpler version of the existing rules by eliminating home team wins or loses: *Watch the soccer game and go out with friends and then have a beer.*

In summary, with decision trees, it is simple to interpret and implement the final results; few assumptions are needed about distributions and linearity; decision trees are not sensitive to outliers, are good for screening purposes, have fast execution and produce results with good performance.

On the other hand, although decision trees are easy to use, there are a number of disadvantages: sometimes, the number of nodes is reduced too drastically, leading to coarse segmentation; there is no hypothesis testing or parameter estimation; and specialised software is often needed to utilise the full range of capability of the method.

Decision trees are particularly useful if the target variable is not continuous but is expressed in classes. Decision trees have the big advantage that they are intended to be read and understood rather intuitively. Using decision trees, the results of analysis are clear and understandable for non-data miners as well as data miners. In addition, the single pathways in the tree can be simply translated into any programme and applied, avoiding an otherwise intricate conversion and implementation process.

Although the concept of decision tree analysis is straightforward, nevertheless, there are a huge number of algorithms available for creating the decision trees. The algorithms differ in the selection procedures of the relevant input variables, the splitting criteria, the number of branches of the tree, the symmetry of the tree and the pruning methods.

Basically, most decision trees depend on a procedure which is called 'recursive partitioning'. Recursive partitioning is an iterative process in which the data inventory is split into smaller and smaller subsets. Besides, these subsets always become more uniform, so more and more observations or customers have the same value of the target variable. In the ideal case, at the lowest or finest level, the target variable in the subset only has a single value. The lowest level is called the leaves (or sheets), and the branching out in between occurs at the nodes (or knots) (see Figure 6.20).

As with almost all data mining procedures, the generalisation and globalisation of the forecast model only makes sense if decision trees are a suitable choice of analysis method. In addition, there is also with decision trees the problem of over-fitting. Here, the model is so differentiated that it considers specific features (e.g. content and statistical outliers) in the analysis data too strongly, and if the model is applied, it could lead to mistakes. One can prevent this situation, by working with a training sample and a test sample.

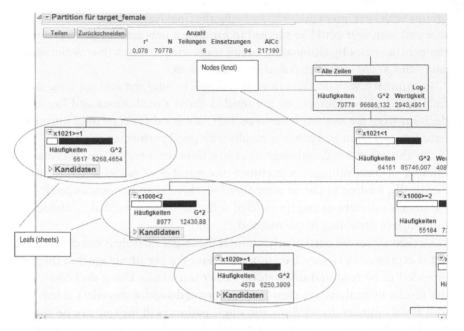

FIGURE 6.20 Decision tree with leaf and knots.

Besides the development of forecast models, decision trees are also used to determine the 'most important' variables (variable selection), so that these variables can be used later in other procedures such as regression analysis.

6.5.2   Selection Procedures of the Relevant Input Variables

The choice of input variables for the decision tree is made on the basis of statistical tests. The chi-square test is commonly used and is often the default setting in data mining software.

The variable with the strongest single influence is taken as the first branch. However, it can turn out in the course of the construction that a combination of 'less important' input variables is better than this variable when they are used in combination. If this is noticed, then the tree should be calculated anew without the 'most important' single variable.

6.5.3   Splitting Criteria

A tree can have two or more branches, and it is important to know how a variable is split and where the division should take place. The algorithm proceeds as follows: for every variable, every possible split version is calculated; the split that is chosen is the one that gives the best discrimination as regards predicting the target variable. In fact, 'makes the best discrimination' means that the likelihood is very low that a second observation in the same side of the split would have the opposite target value. The most common current algorithms for the assessment of the split with a binary target variable are a chi-square, Gini index ($2P1 * (1 - P1)$) and entropy ($P1\log P1 + P2\log P2$) where $P1$ is the proportion of cases with Target$=1$ and $P2$ is the proportion of cases with Target$=0$. In wide areas of statistics, the Gini index is also called the 'Simpson Diversity Index'.

6.5.4   Number of Splits (Branches of the Tree)

It is important for the construction of a decision tree to fix how many branches the tree should have. Often, trees with only two or three branches are chosen as this makes it easier to construct and apply the corresponding rules.

6.5.5   Symmetry/Asymmetry

Decision trees can be symmetrical or asymmetrical. An asymmetrical tree is one in which the number of cases is greater in one branch than in the other branch, whereas in a symmetrical tree, it is a requirement that the number of cases is equal. This naturally affects the positions of the splits in the variables. Normally, one chooses to construct an asymmetrical tree because the different contributions to the separation of the target variable turn out better. Some data mining software like SAS Enterprise Miner, for example, generally calculates asymmetrical decision trees.

6.5.6   Pruning

Pruning means to cut branches to improve results as in horticulture. An automatic pruning option is usually offered by the software, and its aim is to avoid over-fitting. This process is intended to eliminate all areas of the tree which contain only the specific features of the training sample or which contribute non-significantly to the separation in the target variable. Another problem area that can be ameliorated is that automatic pruning may show up variables that cannot be generalised, for example, variables that are important in the training sample but may not be in evidence in the test sample. These sorts of problems can arise as a result of poor sampling or with sparse variables that have very few different values. In Figure 6.21, the software has the option of automatic pruning, or you can do it yourself by clicking on the buttons circled in the top left-hand side of the top. Clicking on 'Teilen' will give a further split, and clicking on 'Zuruckschneiden' will make one pruning operation. This

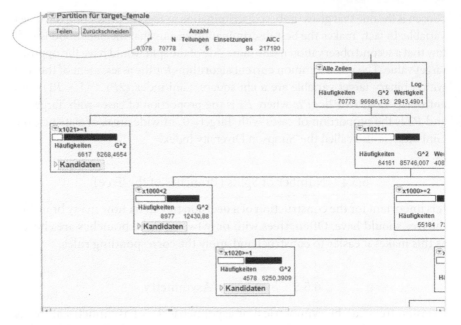

FIGURE 6.21   Decision tree – Pruning.

interactive facility makes tree building very satisfying as it can clearly be seen how the tree builds up.

But note that vigorous pruning does not always make sense.

In some cases, rather big groups which all have the same quality can result from over-vigorous pruning. For example, the aim of using the decision tree is often to select a subset of good prospects from the whole dataset. If the decision tree has just a few very large leaves after pruning, we have to resort to choosing a subset of cases in a haphazard manner which is only a little better than if we had no guidance from the tree at all. It may be better to choose a tree on the basis of the pre-pruned leaves so that we can let the tree help us with choosing the exact number of cases we want by making available the subsets in all the small leaves. With firm productions or similar, it can become a problem if only some of the members of a large group are needed. For example, if you need 30 000 customers for your promotion and the group size is 400 000, how do you choose 30 000? In this case, rather than take the first 30 000, it may be better to use the sub-optimal leaves in place of ordering. In other words, over-fitting can be used as a means of ranking to get a useful sub-group.

As well as the statistical measurement, further evaluation and validation are needed to ensure that the tree model is of high quality and that it will be useful in a general business way.

6.6   NEURAL NETWORKS

Neural networks (also called neural nets) are the data mining method attracting the highest attention, but which actually have the smallest number of operational applications. The application of neural networks to data mining comes from the area of 'machine learning'.

Among the wide range of techniques available in the data mining domain, artificial neural networks are seen as a technique with the capability to produce highly accurate prediction models that can be applied across a large number of different types of problems. Neural networks were originally developed to imitate the function of the human brain when it detects patterns, makes predictions and learns from experience. The knowledge gained from studying brain activities was transferred to computer programmes implementing sophisticated pattern detection and machine learning algorithms to build predictive models from large historical databases.

The two main components of neural networks may be compared to parts of the human brain:

- The node: This corresponds to a neuron.
- The link: This corresponds to relationships between the neurons (axons, dendrites and synapses).

The data analyst has to make many operational decisions before an artificial neural network can be built. These include the number of nodes needed, how the nodes should be connected, what the suitable weights for each link should be and how the training should be conducted.

In order to make a prediction, the neural network receives input data, or predictors, through the input nodes. These values are then multiplied with the weight values stored at the links. A special transfer function is applied in the hidden layer, and the resulting value at the output node becomes the prediction (see Figure 6.22).

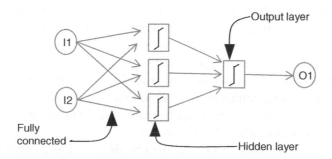

FIGURE 6.22   Artificial neural network.

We can identify the following components of the neural network:

Input layer: Input variable (also transformed input variables).
Units: Every input variable has its own unit.
Output layer: Target variable (also transformed target variable).
Weight: Every input layer is connected by at least one weight with the output layer.

To build a powerful model, an incremental effort is required. In comparison to the other models and algorithms, the examples (historical records) that the neural network learns from are not processed all at once, but are fed in recursively one at a time. Although the neural network learns in a real sense, the algorithm used in this predictive model does not differ much from other algorithms or statistical methods. Indeed, although it is rather a simplification, one could think of neural networks in terms of a combination of regression models or more general multivariate models.

One disadvantage of neural networks is the complexity of the resulting model, which is amplified with the usage of probability and mathematical calculations. Thus, the construction of a model is akin to a 'black box' that requires in-depth analysis. Moreover, as the complexity increases, the predictive model is capable of adding hidden layers of nodes, which further complicates the fine-tuning of the model. There is no doubt that this method is the hardest to deploy for modelling and prediction. Due to its historical origins, the concepts and the names differ markedly from those of statistics. Among other issues, this different naming adds to the mystification of neural networks.

One advantage of neural networks is that all types of data can be analysed with little manual intervention to the input data. Although the method is like a 'black box' with slow execution and with a need for specialised software, neural network models can be automated with no assumptions about distribution and linearity.

Artificial neural network methodology has been used in many facets of business from detecting the fraudulent use of credit cards and credit risk prediction to increasing the hit rate of targeted mailings. It also has a long history of application in other areas such as biological simulations and to learn the correct pronunciation of English words from written text (see Bibliography).

In summary, one can look at neural networks as a combination of interdependent or multistage multivariate models, that is, the models feed into each other. It can be shown that every multivariate model can also be expressed as a neural network as illustrated in Figure 6.23.

Very briefly, a neural network can also be described as a combination of different forecast models feeding into each other. Between the input layer and

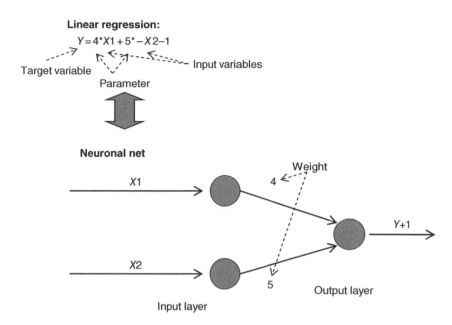

Linear regression:

$Y = 4^*X1 + 5^* - X2 - 1$

Target variable

Input variables

Parameter

Neuronal net

Weight

X1

4

Y+1

X2

5

Input layer

Output layer

FIGURE 6.23   Multivariate model expressed as a neural net.

the output layer, there is nearly always at least one layer or so-called hidden layer. The layer(s) stands for further models feeding in which use the output of the previous layer as an input for the estimate of the target variable. The result of the neural network can be presented diagrammatically as shown in Figure 6.24.

The model is determined according to a multilayer approach. One can distinguish between forward and backward propagation for the development of the model. The most frequent approach is backward propagation (see Figure 6.25). The starting values play an important role. If one looks at a neural network as a combination of different multivariate forecast models, the neural network can be understood theoretically very well, but the results are nearly always very complicated and are not intuitively clear. The much-cited 'black box' refers to the lack of intuitive intelligibility and traceability in the method. This is in contrast to regression modelling and decision trees where it is usually possible to check that the intermediate steps and the results make good business sense.

In practice, neural networks can only process a limited number of input variables; otherwise, the time taken is excessive. In our experience, the number of input variables needs to be limited to approximately 30. Should one want to use neural networks with a much larger number of input variables, then it is advisable to let a decision tree aid the choice of relevant variables beforehand

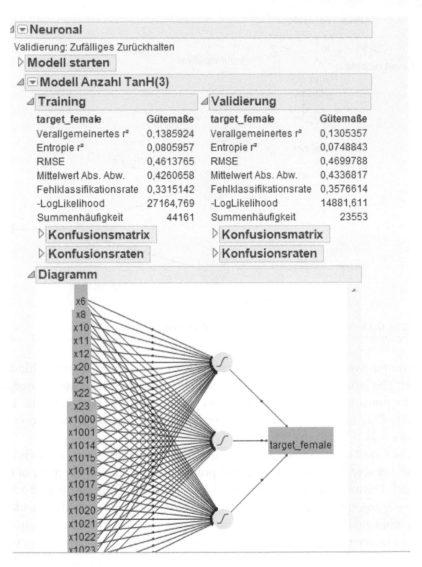

FIGURE 6.24 Neural network model.

FIGURE 6.25 Learning in an artificial neural network through backward propagation.

and then to use the relevant variables of the decision tree as input variables to the neural network.

In summary, the application of neural networks in customer relationship management is rather uninviting because of the poor interpretability and the restriction with the number of input variables as well as the fact that the results of the neural network depend very much on how the learning progresses and which starting values are used in the training stage.

6.7 WHICH METHOD PRODUCES THE BEST MODEL? A COMPARISON OF REGRESSION, DECISION TREES AND NEURAL NETWORKS

We have considered three widely used methods to solve forecasting problems, namely: linear and logistic regression models, decision trees and decision rules and neural networks. Here, we summarise all three methods, giving the advantages and disadvantages that might help to decide which method will be the first choice for solving an actual data mining prediction problem.

Summary: Linear and logistic regression

- Advantages
    - Parameter estimation and hypothesis testing possible
    - Model directly interpretable
    - Execution fast
    - Good for screening purposes
    - Standard software

- Disadvantages
    - Large amount of manual work (to sift through the mass of variables and do preparation to determine the important variables and interactions. Some professional software helps here with pre-selection and the use of stepwise regression)
    - Assumptions about distribution and linearity
    - Sensitive to outliers
    - Performance often not outstanding

Summary: Decision trees and decision rules

- Advantages
    - Simple to interpret and implement results
    - No assumptions about distribution and linearity

- Not sensitive to outliers
- Execution fast
- Good for screening purposes
- Good performance

- Disadvantages
  - Often too few (6–8) final nodes
  - Results of limited value, because not enough groups
  - No hypothesis testing and parameter estimation
  - Needs specialised software

Summary: Neural networks

- Advantages
  - All types of data can be analysed
  - No assumptions about distribution and linearity
  - Good performance
  - Generally applicable predictive equations derived
  - Little manual work

- Disadvantages
  - Difficult to interpret ('black box')
  - Sensitive to outliers in continuous data
  - No hypothesis testing and parameter estimation
  - Execution slow
  - Needs specialised software

6.8   UNSUPERVISED LEARNING

6.8.1   Introduction and Process Steps

There are many alternative methods of analysing a dataset; they vary in the underlying concepts and in the time taken for the analysis. Different methods are appropriate in different circumstances, and in addition, people have their own preferences.

In principle, the approach to unsupervised learning is very similar to the supervised forecast procedures. It follows the data mining process steps as discussed in Chapter 2 and Section 6.2. These steps are revisited here specifically for the unsupervised learning problems. The process steps are:

1. Business task – Clarification of the business question behind the problem
2. Data – Provision and processing of the required data
3. Modelling – Analysis of the data
4. Evaluation and validation during the analysis stage
5. Application of data mining results and learning from the experience

Each of these process steps is described in further detail in the succeeding text.

6.8.2 Business Task

The business task step includes clarification of the problem definition, specification of aims, deciding on the planned application and the application period. A written or oral briefing by the client (industry, marketing department, etc.) is as always very important. Special attention needs to be paid to why the client really wants an analysis. It may be that it is a matter of forming groups on the basis of descriptive characteristics like gender, age and socio-economic class, or it may be that the client really wants to cluster on the basis of buying behaviour. Clustering can be useful whether it is based on the customers or on the products. In practical terms, this means that we can either cluster on cases or variables or both.

An important feature regarding clarifications and applications is to what extent the results of an analysis can be transferred and be generalised for use in practice.

6.8.3 Provision and Processing of the Required Data

To provide the required data for analysis, the following steps must be run through:

- Fix the analysis period, based on the application aim and taking into account the briefing information.

Because we are dealing here with a method of unsupervised learning, it is not necessary to define a target variable.

- Fix the population of customers to be analysed.

This is indicated by the briefing on one hand and on the other hand by the customers' attributes. For example, it may be that the client only wants to consider customers who deal with a particular branch of the company. If the analysis is going to be carried out at branch level, it must be determined whether the customers to be analysed are all customers of that particular branch.

- Generate a working random sample.

Usually there are more than enough customers available with whom to carry out the analysis. Because of this, it is feasible and sensible to work with random samples. A sample size of approximately 30 000–100 000 has been found to be favourable in practice. Evidently, more random samples can be used to replicate the analysis if there are not too many input variables in the analysis. If there are a lot of input variables, one runs the risk that the software cannot process the amount of data in a reasonable amount of time; the processing time can explode in size, and then, there will be an unacceptable waiting time for the results, and they will be less useful in the end.

- Generate input (explanatory) variables.

The input (explanatory) variables can be generated in the usual manner: investigating the data generating individual values from the facts; creating normalised variables; using readily available variables; creating new variables from skilful combinations of existing variables. Remember that all variables should be generated only for the base period.

According to the situation, some of the variables may need to be standardised. This is necessary particularly when the values of the variables have very different meanings, for example, consider 'amount' and 'turnover'; a value of 100 is completely different as an amount than as a turnover. It is necessary to carefully check the data levels and the additional information perhaps available in the data, and the non-information. There needs to be special consideration when the subject is 'unknown'. For example, a case is not male just because the female code has been entered as '0' rather than '1'. Such unknowns can occur where background information is entered automatically.

It is important to decide which variables must be classified and where it is important to complete data and code or recode into dummy variables. According to the procedure to be applied and depending on the desired statements, it is also important to check whether continuous variables would be better classified or not.

There are two ways to approach the classification:

In most data mining software, there are options for data modification and transformation, but usually in these options, you have to work on every variable by hand. The advantage of such a very individual classification is counteracted by the disadvantage of a very high lead time.

With a standard SAS programme, everything can be converted into class variables by dividing the data into sextiles whose class borders are fixed by the programme.

In addition, attention needs to be given particularly to outliers. These are to be removed if necessary or eliminated. A sensible option to fix how missing values will be handled in the procedure is to state that there are 'no missings' in the records.

6.8.4   Analysis of the Data

By analogy with forecasting procedures, the first step is always the descriptive analysis of the data. Then, on the basis of these results and the briefing, the variables are classified if necessary, and appropriate methods and variables are chosen.

The aim is to detect hitherto unknown groupings. There are two main approaches available:

One approach to finding the unknown structures in the data is to carry out cluster analysis or Self-Organising Maps (SOM) and Kohonen networks; the other approach is to use association rule-based algorithms.

The statistical methods are introduced only briefly here to allow an overview. Details of the single methods and their applications are readily available in statistics books (see Bibliography).

SOM and clustering are available in JMP. Should other cluster algorithms be required which are not available in JMP or other data mining software, one can usually find a suitable procedure with the relevant options in the native SAS and if necessary make the SAS code available in JMP.

Here is an overview of clustering, as an example of unsupervised learning:

Enter the input variables to be analysed. Adapt the level of the input variables if necessary, for example, check whether ordinal-level variables are really continuous level. A typical example is the 'amount' variables.

Place all desired variables as INPUT as shown in Figure 6.26. However, notice that there should not be too many variables made available, because the methods generally do not include feature selection algorithms or a check on the number of variables. The computation time increases enormously if there are too many variables. The number of variables can become large particularly if dummy variables have been created for nominal or ordinal variables. Note that from a variable with nominal level, for example, types of product, as many dummy variables are generated as the variable has values, so if a nominal variable has seven possible values, then there will be seven dummy variables. The total number of variables should not be greater than 400–500 according to our experience.

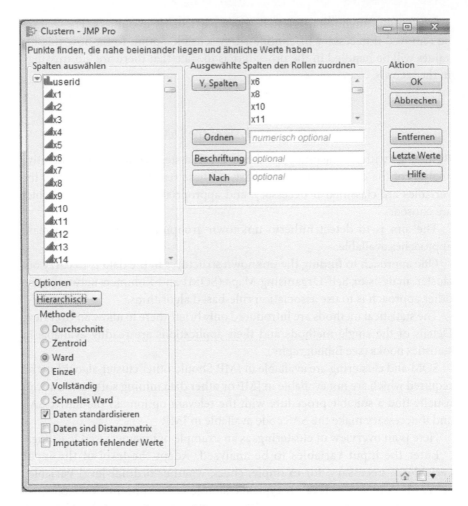

FIGURE 6.26 Basic realisation of clustering Part 1.

Initiate data partitioning, so that a training and test sample are available for the examination of the other procedures. Alternatively, another random sample can be selected with a random sample programme on which the results can be validated when required.

Different methods can be chosen, for example, in Figure 6.27, the clustering method of K-Means has been chosen.

Besides considering the choice of methods, we have to check if there is any default for the sample size used with cluster analysis or SOM or Kohonen networks. A standard random sample size may be built-in and may be highly conservative, such as 2 000; we may prefer to state our own larger sample size, for example, 30 000, which is appropriate for the

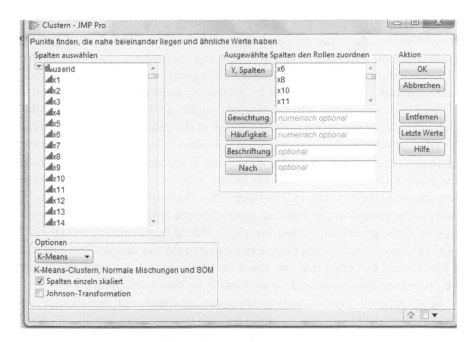

FIGURE 6.27 Basic realisation of clustering Part 2.

particular analysis. However, recall that the larger the sample size, the longer the analysis will take.

According to advance information, one can give the number of the clusters to be found or leave it to the system.

6.8.5   Evaluation and Validation of the Results (during the Analysis)

Evaluation and validation differ in unsupervised learning as compared to supervised learning. In supervised learning, we have a clear model, and the results of applying the model can be checked, for example, by looking at the estimation error. With unsupervised learning, we have no target, only an analysis of the structure of the data. We cannot be sure whether the structure we have found is applicable to the whole dataset or is just an artefact of the sample used in the unsupervised learning. One solution is to carry out a similar style of analysis on both the training and the test samples and see if the patterns found are similar. If the patterns are similar, then we can assume that they are real and are present in the whole dataset.

The process of carrying out an analysis can often be immensely valuable simply because of the focus on collecting a clean, reliable set of data. Sometimes, the benefit of the analysis can be demonstrated in terms of expected financial savings or increased profits.

Another way to check the validity is to compare the results with what is already known about the data and structures. One should always look at the results from the pure business point of view and check that they are reasonable.

Some of the unsupervised methods result in groups (cluster analysis, SOMs), and validation can be carried out by comparing features of the groups in the training and test samples. The comparison can be carried out by looking at the characteristics of important variables in each of the groups, both in the training and the test samples. We can compare the variables in terms of the overall shape of their distribution or by their means and standard deviations. Note that the group numbering in the training sample may not correspond to the group numbering in the test sample. The issue is that the most important variables in each of the groups in the training sample should have a similar importance in one of the groups in the test sample. It should be possible to match each of the groups in the training sample to a group in the test sample. Real data has a lot of noise, and exact matches should not be expected; the comparison can be carried out informally, and we are looking for reasonable similarity.

In operational terms, compare the mean and standard deviation of an important variable, say, age, in group 1 of the training sample with the mean and standard deviation of age in each of the groups in the test sample. We would expect to find a match in one of the groups, and we can then check for matches in the other variables. If the important variables can be more or less matched in this way, then this validates the model. If similar results are not obtained, then it may be better to try a different method of analysis.

6.8.6   Application of the Results

If your data mining tool has an application process available, then that can be used to apply the rules to the old or to a new dataset. Otherwise, you should extract the major rules from the analysis that help you to describe the groups and then apply the rules yourself. For example, group 1 may contain women between the ages of 17 and 22, buying clothes from local boutiques, and so, we can search for these customers in our dataset.

6.9   CLUSTER ANALYSIS

6.9.1   Introduction

The objective of cluster analysis is to gather customers into different groups. Cluster algorithms try to choose the groups in such a way that the structural qualities of the groups are very different to each other whilst being very similar within the groups. Cluster analysis uses all the input variables to 'learn' the structure and determine the groups. The procedures have no selection mechanisms for the input variables, so the analyst should decide which variables are used with the help of business knowledge and pre-analyses. There are various methods of cluster analysis which differ in the way the clusters are constructed, the type of input variables that can be used, the speed of finding the clusters and the way cluster membership is determined.

Cluster analysis can be carried out on the cases or on the input variables.

6.9.2   Hierarchical Cluster Analysis

Clusters can be constructed by building up the sub-groups from the individual customers in a manner referred to as agglomerative hierarchical clustering. Initially, each customer is in their own sub-group; then, the clustering seeks to join two sub-groups and searches for the most similar pair of sub-groups.

The similarity between sub-groups is measured using one of several alternative methods. The centroid method is to minimise the Euclidean distance between the mean values of each input variable in the sub-groups. Euclidean distance is the sum of the squared differences between values. The centroid method assumes that all input variables are measured on a metric scale. Standardisation is recommended to ensure all input variables are on the same scale.

Ward's method is to form clusters in such a way that the total within cluster variance is minimised. The common methods of deciding which clusters join together include:

- Centroid
- Single linkage (also known as nearest neighbour procedures)
- Complete linkage (also known as furthest neighbour procedures)
- Average linkage
- Median
- Ward's method

Similarity between clusters is calculated using all the input variables. This is one of the reasons why the arithmetic time explodes if a very big random sample with a lot of variables is taken as an input file. The clustering proceeds until a stable set of clusters is formed. The agglomeration can be illustrated in a dendrogram which is a sort of tree diagram showing the arrangement of clusters. An alternative faster method is given in most software and is based on the concept of K-Means.

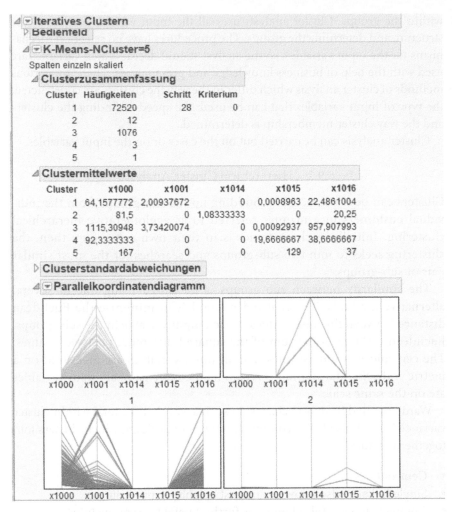

FIGURE 6.28   Realisation of clustering K-Means.

6.9.3   K-Means Method of Cluster Analysis

The K-Means algorithm is based on building up clusters from K initial seed customers or clusters. The value of K may be provided by the software or by the analyst in which case values of 3–10 are not uncommon depending on business knowledge and requirements. The algorithm proceeds by looking at each customer or cluster and deciding which cluster it is closest to and then joining those together. Again, like hierarchical clustering, the distance measure is Euclidean based on differences between mean values of each input variable, and so, standardisation is recommended.

K-Means is a very fast clustering method that can deal with large numbers of input variables and a large number of cases. It is faster than hierarchical clustering because the number of comparisons needed to decide which groups to join together is smaller. A problem with K-Means is that it can lead to very large groups, so that there is not much discrimination available. One solution to this problem is to apply K-Means to the customers in that group only. Another possibility is to increase the number of groups found (i.e. increase the value of K) so that smaller clusters are formed.

6.9.4 Example of Cluster Analysis in Practice

The strength with which the variables have influenced the cluster learning can be seen in the importance column in the 'Results' section; the greater the value, the more significant is the variable. Each software has its own way of representing the clustering results. See, for example, Figure 6.28. Some software provides an interactive option as well.

As can be seen in Figure 6.28, it can happen that just two out of the five clusters have a relevant size. This might lead to a new setup of the clustering, for example, by including or exchanging variables. If the result with just two clusters still appears, it may be that only two clusters exist.

If one is interested in describing the clusters, one can calculate average values and other statistics for every variable per cluster. For dummy variables and O/1 indicator variables, this average value can also be interpreted as a percentage value.

An alternative representation of the results is the tree diagram or dendrogram. This is available in the 'Results' section.

6.10 KOHONEN NETWORKS AND SELF-ORGANISING MAPS

6.10.1 Description

Kohonen networks are an unsupervised learning technique which uncovers structures in data. They are also called SOMs. SOMs can be understood as neural networks for clusters. However, the training methods clearly differ from those of neural networks and forecasting methods. In addition, not only is the output layer a classification variable with few levels (i.e. classes), but the classes are represented in a two-dimensional Kohonen layer.

Each input variable is represented in the input layer by a neuron (or node). The structure of the Kohonen (output) layer is given by the user. A two-dimensional or three-dimensional arrangement supports the later visualisation of the data.

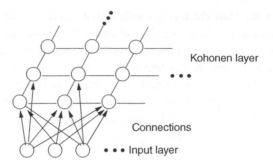

FIGURE 6.29 Kohonen network with two-dimensional arrangement of the output neurons.

The picture in Figure 6.29 shows a Kohonen network with a two-dimensional arrangement of the output layer. Every unit of the input layer is connected with all the units of the Kohonen layer, and a weight vector is assigned to every Kohonen neuron related to the size of the input vector.

The Kohonen neurons are connected with each other. A similarity measure like the Euclidean distance compares the input vectors. The neuron with the greatest similarity to the input layer wins. The weighting of the winner's neuron in the input layer is so modified that the similarity rises further. So far, the Kohonen network works like any other clustering procedure. To generate the topological structure, the weight vectors of the neighbouring neurons of the winner also change. This requires a definition of neighbourhood which is delivered by different functions as, for example, the region function or the cosine. These functions deliver a measure of the distance of every neuron in the Kohonen layer of the winner's neuron which influences the intensity of the weight change. The closer a neuron is to the winning neuron, the stronger his weight vector is adapted. The vectors of nearby neurons are shifted therefore always in similar directions. In this way, clusters containing similar samples are created.

By analogy with cluster analysis, the pre-set random sample size must be increased with SOM analysis.

6.10.2   Example of SOMs in Practice

Under 'Cluster' instructions, one can choose between the different methods. If one chooses SOM, one can also choose the number of rows and columns for the map. In Figure 6.30, three clusters are selected with one row and three columns. The choice of 1×3 clusters is based on previous experience.

The SOM for 3×2 clusters is shown in Figure 6.31. Note that only four clusters are actually used by the method because the other clusters were not needed to explain the patterns in the data.

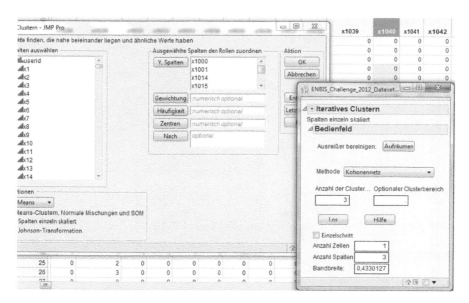

FIGURE 6.30 SOM/Kohonen network realisation Part 1.

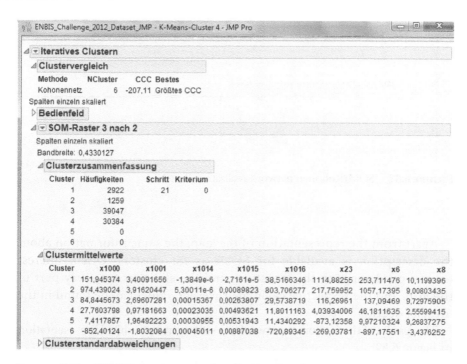

FIGURE 6.31 SOM/Kohonen network realisation Part 2.

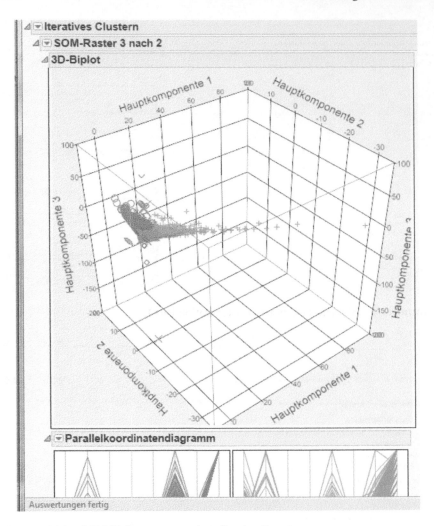

FIGURE 6.32 SOM/Kohonen network realisation Part 3.

Apart from the representation of the map, the same information about the description is available for SOMs as for clusters. Often, interest focuses on describing the structures and the statistics for every part of the map, and the respective calculated average values can be found in the 'Results' section.

The results of the Kohonen network are shown in a graphical representation in Figure 6.32.

6.11   GROUP PURCHASE METHODS: ASSOCIATION AND SEQUENCE ANALYSIS

6.11.1   Introduction

Group purchase methods analyse what products, services or activities customers do together. It is also referred to as market basket analysis. Typical areas of application are the subjects addressed by questions such as:

- What do the customers buy together or in addition?
- Which additional offers can one present to these customers (cross-selling)?
- How high is the chance that the customer shops in two or more industries?

It has to be carefully distinguished at what level the subject is defined as 'together'. The following are possible:

- Together at a cash desk
- Together during a visit
- Together on one purchasing day
- Together in 1 month (or week, quarter, etc.)
- Together within 1 year
- Together within an agreed number of days (e.g. 100 days, 365 days)

For sequence analysis, typical areas of application are the subjects addressed by questions such as:

- Which way does the customer proceed through the store, that is, the route taken through the shop could be vegetables, then stationary, etc.; what does the customer bypass?
- Which product initiates the purchase of another one?
- Which goods are purchased together with campaign or promotional or specially promoted products?

These questions are all related to the order in which actions take place, and so, the order should be included in the data and be accounted for in the analytics.

If no other information is available, purchase day by customer might be a good option for the grouping factor.

If interest focuses on which products the customers take over a longer period of time at the same company, then a summary period of about a year may be preferable. If you have access to data from different companies, then it can be combined to give a richer profile; however, normally, confidentiality and data protection laws usually mean that only 'your' own data can be analysed.

Using a summary period of a year, one can be sure that all seasonal influence can be considered. A longer term period also gives information to what extent single ranges have an attractive strength on others and whether it is possible to develop statements like: X% of the buyers of product A are also buyers of product B, the classical cross-selling.

To provide the required analysis data, the following steps must be run through:

1. Fix the analysis period:

The analysis period must be fixed based on the application aim and taking into account the briefing information. Special notice must be taken of the summary problems described in the area of application.

2. Fix the population of customers to be analysed:

The basic population of customers to be analysed is determined by a combination of the business briefing and the qualities required from the customers or the branch to be analysed. Should the group purchase analysis be carried out at branch level, it must now be fixed whether to include all customers of a branch or only specific customers.

3. Define the aggregation level of the data and the receipts to be analysed:

To ensure that the group purchase analysis gives informative results, it is necessary to consider the aggregation level carefully. The main issues are:

*   Does this analysis make sense at the article (order detail) level?
*   Which level (department, industry, usage) offers a result that can be generalised? It is more problematic to express results at a more detailed level; the possible combinations may not be easy to replicate and cascade because orders at some levels of detail may no longer appear in the next season.
*   Which time/invoice aggregation is required for the analysis (e.g. at purchasing day level or at annual level)?
*   Are the results relevant for single customers or do they extend to customer groups, such as households?
*   What is my grouping level (receipt, customer, household, company) for the analysis?
*   What criteria do the receipt holders follow, for example, everybody uses at least one voucher from multimedia? What does this mean for the interpretation of the data?
*   Is the order of the data/transactions important?

Basically, it is valid to say that all relevant data for the analysis follows the picture 'long and narrow', that is, a receipt has several rows in the file which can be

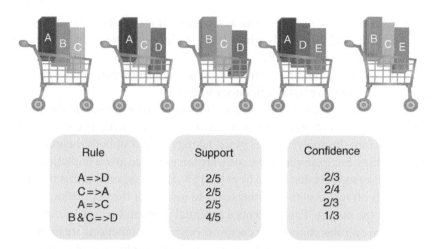

Rule	Support	Confidence
A=>D	2/5	2/3
C=>A	2/5	2/4
A=>C	2/5	2/3
B&C=>D	4/5	1/3

FIGURE 6.33 Association rules.

summarised about a common key (e.g. customer's number or customer's number and purchasing day). These kinds of analysis can only process an input variable, an ID/Target and a sequence variable.

6.11.2 Analysis of the Data

Group purchase analysis can only be carried out using specialist data mining software. Most of the software works with an interactive screen where input can be defined and the style of output can be specified. All data preparation should be carried out before the tool is used as quite often the tools do not include options for manipulating the data. There are usually options to carry out the analysis with or without consideration of the order of the transactions in the receipt.

For many analyses of this kind, the time order is not decisive, or the date given when material was stored is not reliable, particularly if these analyses are carried out at industry or usage level.

An association rule, R, is a statement of the form $A \rightarrow B$ where A is the premise of the rule and B is the consequence. The premise and consequence must be disjoint in the sense that they cannot both occur at the same time.

The rules extracted by the analysis give us general information about the associations implied in the data. The quality of the rules can be evaluated by their support, confidence and lift (see Figure 6.33).

Support is defined as the proportion of cases in which the association occurs divided by the total number of receipts (i.e. opportunities). Support is a measure of how frequently the rule occurs.

Confidence is defined as the proportion of cases in which the association occurs divided by the number of cases containing the premise.

Lift indicates the strength of the association. It is calculated by the ratio of conditional probability of B given that A has occurred divided by the unconditional probability of B.

Note that support and lift are the same in an association rule regardless which way round A and B are but confidence can differ.

Different software allows different restrictions to be imposed on the way the analysis is carried out and the way the output is presented. These have to be specified before the analysis takes place, taking account of both theoretical and business issues. For example, if the software forces you to give a limitation on the output such as support that has to be at least 5%, then it is possible that important information on rare events that may apply to just 1% of the population may not show up in the output. This presents a potential problem because in many businesses the main rules concerning customer buying combinations are known and are quite stable but the special focus is just on the more obscure rules applying to a small % of customers. Therefore, take care to ensure that business needs are likely to be met and that it is possible to generalise the results for future actions.

6.11.3   Group Purchase Methods

There are two main approaches to consider:

- Association analysis (undirected)
- Sequence analysis (directed, i.e. considering the purchase order)

A range of different algorithms are available to generate rules for associations and sequences. Some of the algorithms date from the nineteenth century, however, and are applicable today only because of the prevalence of powerful computers. A common feature of all rules is that they relate one element of the receipt to one or more other elements. Confidence in the rule indicates the likelihood that the other items appear because of an actual association rather than just by chance.

Therefore, association analysis is based on the theory of probability. The patterns found are described with association rules giving just the association if the undirected approach is appropriate and rules for a sequence of associations if the directed approach is required.

There are some interesting applications of sequence analysis especially in web analytics and social science contexts; see Bibliography.

6.11.4   Examples of Group Purchase Methods in Practice

In Figure 6.34, you will see two different sample datasets. Note that a unique key for grouping is required to do the analysis. At the left-hand side, the unique key is named Auftrags_num (order_id). If this key is used, you can analyse what kind

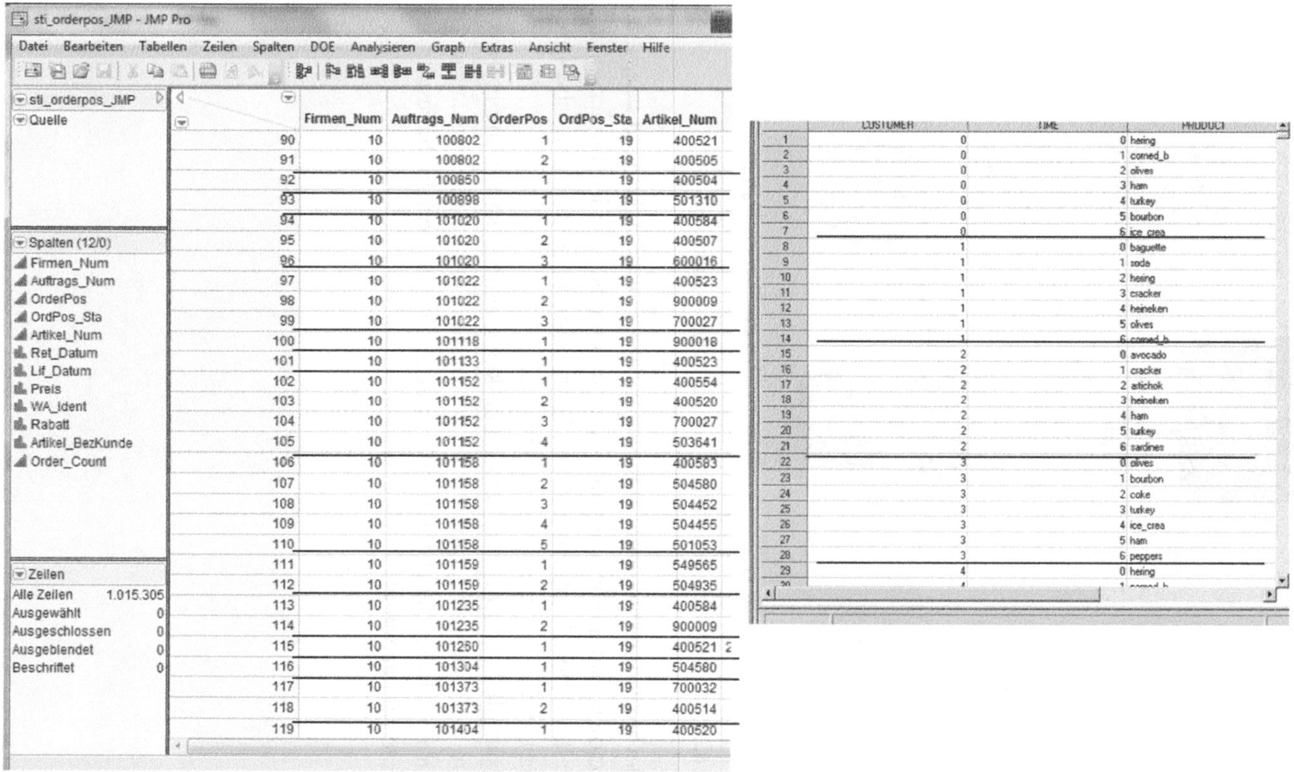

FIGURE 6.34 Example data.

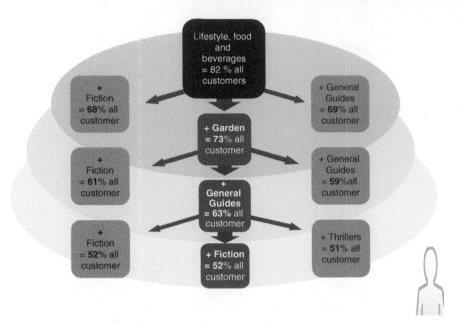

FIGURE 6.35 Results representation (market basket analysis).

of articles are brought together at the same time. On the right-hand side, you see a well-known example where you can group on customer level independent of whether it was bought together or not. Both examples contain a variable (order_pos, time) that indicates the shopping order.

An example of results from market basket analysis is shown in Figure 6.35.

Figure 6.35 shows a graphical representation of several association rules found in a publisher dataset. It provides a good way to explain the results of the analysis to management.

7

Validation and Application

7.1 INTRODUCTION TO METHODS FOR VALIDATION

Data mining can be used to explain the relationships between variables. It can also be used to produce models that can predict outcomes for each person or case. The closer the predictions are to the observed outcomes, the better the model. The model is assessed in three stages: business evaluation, statistical validation and application on the full population including the corresponding target variables.

A Practical Guide to Data Mining for Business and Industry, First Edition.
Andrea Ahlemeyer-Stubbe and Shirley Coleman.
© 2014 John Wiley & Sons, Ltd. Published 2014 by John Wiley & Sons, Ltd.
Companion website: www.wiley.com/go/data_mining

The importance of business evaluation has been stressed throughout the book and refers to checking all the time that the models make sense in the business scenario. As discussed in Section 2.6, evaluation and validation of the models are carried out during the analysis stage. Two or three of the most promising models are selected from the contending models and applied on the test samples. The statistical validation and results can then be compared to help choose one or more final models.

The chosen models are then applied to the full population, and there is a further occasion when it is important to assess the results of the model. This is the final and very important stage of validation.

Validation methods include tests on the quality of the model. However, in data mining, sample sizes are typically so large that most goodness-of-fit tests will fail because the large sample size makes the tests very sensitive to tiny discrepancies. For example, if we try to test a dataset of 10 values for Normality, then the null hypothesis is not likely to be rejected simply because there is not enough evidence in 10 points to reject Normality. However, if we test the null hypothesis of Normality with a dataset of 10 000 values, then we may well find that the null hypothesis is rejected even if the data appear to be Normal because the large sample size makes the test very sensitive to even the smallest deviation from Normality. In the same way, a goodness-of-fit test for a data mining model may be too strict in a practical sense. We therefore make use of other ways of validating the models such as lift and gain charts, sensitivity, the confusion matrix and Receiver Operating Characteristic (ROC) curves.

7.2 Lift and Gain Charts

Lift and gain charts are a convenient way to assess the validity of the model on a visual level. These charts show the relationship between observed behaviour and that predicted by the model. A cumulative gains chart can also be used.

In both charts, the data cases are ordered by expected behaviour as regards the target. So if one person is predicted to buy with a high probability, then they will appear high up in the ranking as compared to a person with a lower probability. For each person, we know the actual behaviour because this is given by the target variable. For individuals, the target may be classified as 0 or 1. Normally, the cases are grouped so that the chart is smoother. The group response can be presented as a %, that is, the % of people in a particular group who bought the item in question.

In a gain chart, the actual % response is on the vertical axis, and the order percentile of the expected behaviour of the sample is on the horizontal axis.

The first 10% on the X-axis represents the group with the highest affinity to buy. If the model is good, then the observed % response should be high for the group with the highest predicted response. The % response will then decrease from left to right as the predicted responses decrease.

Most of the time, the data is grouped into batches of 10% of expected behaviour, so there are 11 points on the chart. In a decision tree, if the leaves represent smaller than 10% of the dataset, then they would be combined, and if a leaf is more than 20%, then there will points on the same level (see Figure 7.1).

In a lift chart, the data is ordered as before, but the vertical axis now shows the improvement in response over that in the general or sample population. If you have a binary target variable and your sample is 50% in each category, then in the sample, the overall % response is 50%. If your observed response in the top 10% of customers is 80%, then the gain chart would show 80%, but the lift chart shows 80%/50% which is equal to 1.6. In other words, the lift chart shows that your model has lifted the response in that group by a factor of 1.6 (see Figure 7.2).

The lift and gain charts can be used to compare models. Generally speaking, a model is good if its lift and gain charts show high values on the left hand side of the plot before becoming lower on the right hand side. However, usually, we

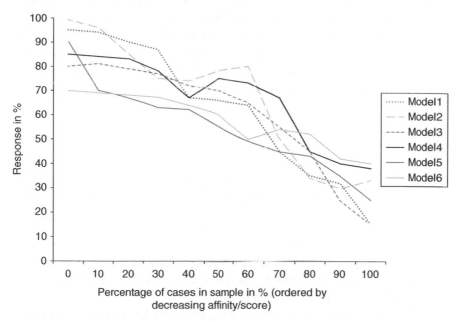

FIGURE 7.1 Gain chart to compare models.

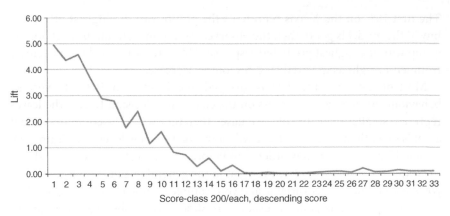

FIGURE 7.2   Typical lift chart (on full population).

are interested in particular parts of the charts rather than the whole picture. For example, we might want a model that is good at predicting the top 20% to offer them VIP handling or a loyalty award. In this case, we are not so concerned about the right-hand side of the graph with the lower potential customers but just with the left-hand part with the top 20%.

Lift and gain charts can be used with any type of target variable although they are more commonly used with binary target variables. If the target variable were expenditure, for example, then the customers would be ordered by expected expenditure, and the gain chart would show the observed expenditure on the vertical axis, and the lift chart would show the expenditure as a % of average expenditure.

7.3   MODEL STABILITY

We can assess the stability of the model by examining it using a lift chart with 1% scaling. The fluctuations in the chart should be low particularly in the planned decision area (see Figure 7.3).

It is important to make sure that the results are also stable when applied to the whole population and have good prediction ability. The favoured model is applied to all customers having been built with the data from the analysis period (base and target period). The predicted results for all customers are compared to the known real target results.

Technically, this can be carried out using the 'score' knots in the data mining software as well as by a script code in SAS or another programming language.

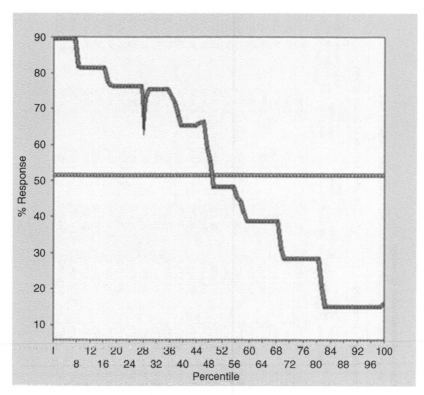

FIGURE 7.3   Example for a model with one unstable area.

In terms of the working time and arithmetic time required, the normal SAS programme is to be preferred. In addition, the score codes from the elective model are generated in the standard SAS programme and can be analysed using any other software, for example, they can be inserted into an Excel spreadsheet for further assessment of the model (see Figure 7.4).

In Figure 7.4, the predictive model has been used to identify 20 groups of customers. The real results for these customers are listed in the central columns of the spreadsheet. A response factor of 0.3 is used to estimate the expected revenue from each group. The predictive model makes it possible to group the customers according to their likelihood of buying so that the top groups can be selected for promotional effort. Without the model, it would not be possible to identify the most likely buyers, and the expected revenue from marketing to a subset of say 20% of customers would just be 20% of the average results over all customers including those likely to buy and those who are unlikely to buy. Using the predictive model ensures that the expected revenues are much

Scoring results: determination of edition

Assumption: Costs per unit 0,70 €, adjustment factor response 0,3, estimated revenue per customer 12 €

		Rating after scoring		Real results of the analysis inventory				Expected results of future action			
	Group	Number of Customers	Number of Customers cumulated	Number of buyers in departments	Buying Rate	Cumulated Buying Rate	Revenue (analysis period in departments)	Estimated cost per group	Expected buyers adjusted by factor	Expected Buying Rate adjusted by factor	Expected revenue per group (real av. Rev.. Buyer * exp. Buyer)
Very good	1	50.000	50.000	20.604	41,20%	8,25%	328.322,0	35.000	6.181	12,36%	74.174,40
	2	50.000	100.000	16.721	33,44%	14,95%	246.750,0	35.000	5.016	10,03%	60.195,60
	3	50.000	150.000	10.766	21,53%	19,26%	152.848,0	35.000	3.230	6,46%	38.757,60
	4	50.000	200.000	9.107	18,21%	22,91%	132.450,0	35.000	2.732	5,46%	32.785,20
	5	50.000	250.000	9.025	18,50%	28,52%	125.896,0	35.000	2.708	5,42%	32.490,00
	6	50.000	300.000	8.933	17,87%	30,10%	127.080,0	35.000	2.680	5,36%	32.158,80
	7	50.000	350.000	8.007	16,01%	33,31%	112.690,0	35.000	2.402	4,80%	28.825,20
	8	50.000	400.000	8.134	16,27%	36,57%	117.566,0	35.000	2.440	4,88%	29.282,40
	9	50.000	450.000	8.156	16,31%	39,83%	115.885,0	35.000	2.447	4,89%	29.361,60
	10	50.000	500.000	7.209	14,42%	42,72%	100.967,0	35.000	2.163	4,33%	25.952,40
	11	50.000	550.000	6.624	13,25%	45,38%	94.278,0	35.000	1.987	3,97%	23.846,40
	12	50.000	600.000	5.516	11,03%	47,58%	67.026,0	35.000	1.655	3,31%	19.857,60
	13	50.000	650.000	5.496	10,99%	49,79%	66.916,0	35.000	1.649	3,30%	19.785,60
	14	50.000	700.000	5.657	11,31%	52,05%	69.432,0	35.000	1.697	3,39%	20.365,20
	15	50.000	750.000	5.543	11,09%	54,27%	67.842,0	35.000	1.663	3,33%	19.954,80
	16	50.000	800.000	5.585	11,17%	56,51%	69.919,0	35.000	1.676	3,35%	20.106,00
	17	50.000	850.000	5.262	10,52%	58,62%	63.706,0	35.000	1.579	3,16%	18.943,20
	18	50.000	900.000	5.380	8,76%	60,37%	60.561,0	35.000	1.314	2,63%	15.768,00
	19	50.000	950.000	4.429	8,86%	62,14%	60.447,0	35.000	1.329	2,66%	15.944,40
Very bad	20	4.616.931	5.566.931,00	94.511	2,05%	100%	1.137.509,0	3.231.852	28.353	0,61%	340.239,60
	Gesamt	5.566.931		249.665	4,48%	100%	3.318.060	3.896.852	74.900	1,35%	898.794,00

Recommended edition

FIGURE 7.4   An example of model control in Excel.

higher. This advantage applies regardless of the response factor used. The value of 0.3 is a rule of thumb and includes the effects of process and operational issues. It is important to include this factor if the estimates are going to be reported to management; otherwise, there will be unrealistic expectations.

The main point here is that using the predictive model improves the expected revenue; looking at Figure 7.4 you can target those expected to give high revenue and ignore those who are unlikely to buy. Also you can calculate the advertisement costs beforehand, and so with the expected buying rate and revenue, you can predict the cost per order and decide whether this cost fits within your strategy and hence decide whether to go ahead or not with a campaign. If you decide to go ahead with the campaign, the groups obtained from the predictive model can help you decide how many people to contact because the groups clearly show where there are cut-offs in added value.

7.4   SENSITIVITY ANALYSIS

The model learnt on the training sample is applied to the test sample and the validation sample. The prediction accuracy should be similar in all the samples. If the target variable is binary, then the results are set out in a table comparing predicted results with observed results (considered in this context to be true results). The table can be summarised in terms of the sensitivity and specificity. We want high values of both.

Figure 7.5 shows some example data of the number of correct and false predictions for each target outcome, and these figures will be used to define the terms.

Taking Target = 1 to be the positive outcome and Target = 0 to be the negative outcome:

Sensitivity is defined as the number of correct positive outcome predictions divided by the number of true positive outcomes. This is $5040/(5040 + 960) = 5040/6000 = A/(A + B) = 84\%$.

	Predicted target = 1	Predicted target = 0	Total
True target = 1 (positive)	$A = 5040$	$B = 960$	$A + B = 6000$
True target = 0 (negative)	$C = 1080$	$D = 4920$	$C + D = 6000$
Total	$A + C = 6120$	$B + D = 5880$	$A + B + C + D = 12000$

FIGURE 7.5   Prediction versus outcome.

Specificity is defined as the number of correct negative outcome predictions divided by the number of true negative outcomes. This is $4920/(1080 + 4920) = 4920/6000 = D/(C + D) = 82\%$.

Sensitivity and specificity are related to other terms and concepts including:

- False positive rate (α) = type I error = $1 -$ specificity = $C/(C + D) = 1080/6000 = 18\%$
- False negative rate (β) = type II error = $1 -$ sensitivity = $B/(A + B) = 960/6000 = 16\%$
- Power = sensitivity = $1 - \beta$
- Likelihood ratio positive = sensitivity/$(1 -$ specificity$) = 84\%/(1 - 82\%) = 4.7$
- Likelihood ratio negative = $(1 -$ sensitivity$)/$specificity = $(1 - 84\%)/82\% = 0.2$

Note that if the population has a higher proportion of people or cases with Target = 0 as often occurs in marketing where 95% of cases having Target = 0 is not uncommon, then the specificity is likely to be high as a lot of people will be correctly estimated as having Target = 0, but sensitivity may be low because there are only a few people in the Target = 1 category and getting the prediction wrong for a few people will have a large impact. For example, there may be 9 500 Target = 0 people and only 500 Target = 1 people in a population of

Name	Importance	Role	Rules	Variable Label
ABC_WERT	1.0000	input	1	WERT_VOR_1Q
ET_BR_0094	0.6980	input	1	
M_GES	0.5446	input	2	
ET_BR_0094_3	0.5445	input	3	
ANZTAGE_LTZTKAUF_ABT_015_2	0.5094	input	1	
DB_BR_0094_3	0.4427	input	1	
M_GES_REN_CO_DWB_4	0.4100	input	1	
M_CO_DWB_2	0.3617	input	1	
DB_BR_0094	0.2369	input	1	
ET_BR_0094_2	0.2361	input	1	
ET_KF_0001_4	0.1827	input	1	
ANZTAGE_LTZTKAUF_BR_0005	0.0595	input	1	
DURCH_UMS_ET_BR_0033_4	0.0335	rejected	1	
ANZTAGE_LTZTKAUF_ABT_178_2	0.0335	rejected	1	
ANZTAGE_LTZTKAUF_ABT_005	0.0000	rejected	0	
ANZTAGE_LTZTKAUF_ABT_105	0.0000	rejected	0	
ANZTAGE_LTZTKAUF_ABT_058_4	0.0000	rejected	0	
ABC	0.0000	rejected	0	KZ_VOR_1Q
ANZTAGE_LTZTKAUF_ABT_057_2	0.0000	rejected	0	
ANZTAGE_LTZTKAUF_ABT_168	0.0000	rejected	0	
DB_BR_0035	0.0000	rejected	0	
ANZTAGE_LTZTKAUF_ABT_177_3	0.0000	rejected	0	
ABC_2	0.0000	rejected	0	KZ_VOR_2Q
ANZTAGE_LTZTKAUF_ABT_011_4	0.0000	rejected	0	
ANZTAGE_LTZTKAUF_ABT_206_3	0.0000	rejected	0	

☑ Export role as indicated in the table

FIGURE 7.6 List of the variables relevant for the model.

10 000. If 100 people are misclassified in each direction, then 100 out of 9500 Target = 0 people are wrongly predicted to buy and specificity is 9400/9500 = 99%. However, if 100 out of 500 Target = 1 people are wrongly predicted not to buy, then sensitivity is only 400/500 = 80%.

Sensitivity analysis also has a wider meaning; it is the study of how the uncertainty in the output from our model is related to different sources of uncertainty in the input variables. Data mining software provides a list of the variables relevant for the model as shown in Figure 7.6.

It is important to know how the predictions are affected by uncertainty in key variables, for example, if a model includes age, how sensitive are the predictions to the uncertainty in the age measurement we are using? Using binning helps to reduce the sensitivity because the input variables are grouped in such a way that variation such as 45.6 years old instead of 46.1 years old will not affect the results (unless the discrepancy causes the age to cross into a different bin).

7.5 THRESHOLD ANALYTICS AND CONFUSION MATRIX

The confusion matrix is a special version of the prediction versus outcome table discussed earlier. The threshold for the two categories has to be determined and affects the values in the confusion matrix.

The predicted outcome for a customer from a model may be a probability such as 80%. Does this classify the customer as someone who will buy or not? What is a suitable discrimination threshold for the probability? There may be a natural breakpoint for the predicted outcome, for example, 50% because everyone below is less likely to be in the target group than not. However, we may only be interested in those with a very high target probability in which case it would be better to make the threshold 85%. Threshold analytics is based on experience and other business intelligence.

In Figure 7.7, the threshold for 0 and 1 is at 0.5, and the results are given as figures and row percentages for the model learnt on the training sample and applied to the training sample and then to the validation sample.

Using these figures, the sensitivity in the validation sample is 4860/(4860 + 1140) = 4860/6000 = 81%. This compares favourably with the sensitivity in the training sample calculated for Figure 7.3 which is 84%.

Note that all of the four values in both of the tables are similar, which is good. A slight difference may be acceptable, but it is undesirable to have a model with a big difference between results when the model is applied to the training and validation samples.

Train		To (predicted target)		
		1	0	all
	1	**5040**	960	6 000
From (real target)	0	1080	**4920**	6 000
	all	6120	5880	12 000

Validation		To (predicted target)		
		1	0	all
	1	**4860**	1140	6 000
From (real target)	0	1020	**4980**	6 000
	all	5880	6120	12 000

FIGURE 7.7   Confusion matrix for comparing models.

7.6   ROC CURVES

A ROC curve is a plot which illustrates the performance of a binary variable as its threshold is varied. It is created by plotting the sensitivity versus (1 – the specificity) at various threshold settings. The ROC curve therefore illustrates the trade-off between true positive and false positive. True positive (sensitivity) is a beneficial outcome, and false positive (1 – specificity) is a costly outcome. Note that the curve passes through (0,0) when the threshold for being positive is one so that nothing counts as positive, and (1,1) when the threshold for being positive is zero, then everything counts as positive. The ROC curve for a poor model would be an approximately straight line connecting these two points. In contrast, a good model has a well-rounded curve almost to the extent of being left angular (see Figure 7.8).

The proportions of people with each target value influence the interpretation of what ROC curve indicates a good model. If in real life you have 95% with Target = 0 and 5% with Target = 1 and you use the original proportions in sampling, then it is easy to get good estimation of the Target = 0 people, and so the ROC curve will have the appearance of a good model as more people are correctly estimated as Target = 0 and specificity is high and the point is above the 45° line. However, if you are sampling this population with a 50/50 sample, it is unlikely

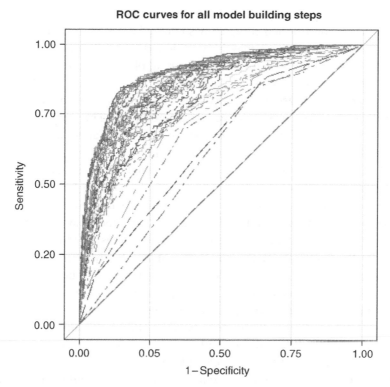

FIGURE 7.8 ROC curve development during stepwise predictive modelling.

with real data that you get a rounded ROC curve, and you are more likely to get a flattish line. If the population and the sample are both split 50/50 between Target = 0 and Target = 1, then the more round the ROC curve, the better the model.

ROC analysis can be used to help select the best models, as a good model will give points above the diagonal line representing good classification results (better than random). This is demonstrated in Figure 7.8 which shows progressively improving models from stepwise regression. The choice of models is usually carried out independently from cost considerations, but ROC analysis can also be used to compare models with similar costs, and this leads to sound decision making.

7.7 CROSS-VALIDATION AND ROBUSTNESS

If a model is built from a training sample and then applied to the training sample, the confusion matrix gives an optimistic picture as the model is tuned to the data used to build it. Data partitioning is used when there is enough data that separate samples can be used for training and testing and validation. The dataset may be partitioned completely so that all cases are included in one of the samples, or the dataset may be partitioned and then samples are taken within each partition. When the models are applied to the test or validation samples, this is referred to as true validation.

However, if the dataset is not very large and there are not enough cases for true validation, then cross-validation is an alternative. There are a number of variations including the 'leaving one out' method. This alternative is to build the model on the dataset leaving out one (or more) case(s) and then apply the model to the cases left out. The procedure is carried out repeatedly, leaving out a different observation each time. The confusion matrix is built up from the accuracy of the results of applying the model to the cases left out.

Cross-validation uses more computation time and may still not be a trustworthy way to validate the model. If the model is fitted on all the data except for the cases left out, then it is fitting it too closely to the data available at that time. That may be acceptable in some data situations, but in marketing, the data is moving on all the time, for example, a change in marketing, product mix or business focus by competitors may have a knock-on effect on your business. Using the whole population except for the one or more cases left out can introduce the danger of over-fitting to a specific data situation. Where there is sufficient data, then it is preferable to learn on one sample and test the model on one or more other samples. These samples could even be from different time periods which would give a more stringent validation test. This practice has been found to produce more robust models.

7.8   MODEL COMPLEXITY

Models are valuable if they have good forecast ability, and their mistake rate as indicated by the confusion matrix (in training and validation) is low. In addition, it is important in practical data mining to make sure the models are plausible and useful. One issue is the model complexity. According to the type of problem being addressed, it can make sense to prefer models that use a large number of variables. This is contrary to the usual statistical practice of preferring models with a lower complexity (i.e. as few relevant variables in the model as possible). However, in the application of the forecast models, it can make sense to obtain a very detailed ranking of the customers. This requires that each group of customers identified contains a reasonably small number of customers. This satisfactory situation is more likely to be obtained if the model contains many variables which will contribute to the segmentation of the customers by the model.

PART III

DATA MINING IN ACTION

8

Marketing: Prediction

A Practical Guide to Data Mining for Business and Industry, First Edition.
Andrea Ahlemeyer-Stubbe and Shirley Coleman.
© 2014 John Wiley & Sons, Ltd. Published 2014 by John Wiley & Sons, Ltd.
Companion website: www.wiley.com/go/data_mining

The recipes in this chapter are about searching the customer base with a view to targeting customers for promotions and predicting outcomes.

The concept described in Recipe 1 appears in many different guises. The applications differ, and the similarities to Recipe 1 are not always easily recognised. In Sections 8.2–8.9, we give typical examples where the variations might arise. In these variations, the methodology in Recipe 1 can be followed up until the implementation. In the following sections, we give typical examples for each application and the special considerations required for the implementation.

8.1 RECIPE 1: RESPONSE OPTIMISATION: TO FIND AND ADDRESS THE RIGHT NUMBER OF CUSTOMERS

In this recipe, the aim is to find a subset of customers to contact. Rather than contact the whole database with a promotion or sales campaign, it is beneficial to use data analytical techniques to identify the optimum subset and avoid wasted resources in contacting customers who we have evidence to suggest have little or no interest in the campaign. Hence, the scientific approach is used to save money, time and effort.

Industry: The recipe is relevant to everybody using direct communication to improve business, for example, mail-order businesses, publishers and online shops. Department stores and supermarkets are included if they offer loyalty cards so that they know something about individual customers.

Areas of interest: The recipe is relevant to marketing, sales and online promotions.

Challenge: The challenge is to find and address the right number of customers to optimise the Return On Investment (ROI) of a marketing campaign (e.g. direct mailing) in a stable set of data (i.e. data that are complete to the date in question). The requirement is for cost reduction by reducing the number of customers contacted in a marketing campaign. The aim is to create a balance between the cost of offering products and the disadvantage of losing sales.

Typical application: The recipe is relevant to a new car model. The automotive company has over 5 000 000 customers on its books and wants to target the 10 000 customers most likely to be interested in the new model. A sample size of 10 000 is chosen because this fits with the promotional budget and the past evidence that % take-up is 5% so that extra expected sales from the marketing campaign are about 500 vehicles (see Figure 8.1).

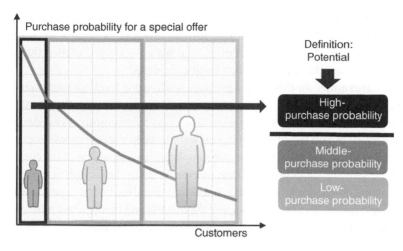

FIGURE 8.1   Selecting the subset of customers with high purchase probability.

Necessary data: The areas of data that could be used must have some direct relationship to the customer reactions or must have come directly from the customer (e.g. data directly from the purchasing process or marketing activities or application form, e.g. for a loyalty card).

Population: The population is defined according to the problem and the briefing. It has to be discussed whether all customers are included (active and inactive ones) or just those belonging to a relevant campaign. Note that campaigns can be highly seasonal in which case we need to consider the population for at least one cycle, for example:

1.   We are considering a Christmas campaign, so for our sampling frame, we consider all customers who are defined as active customers in the company from October the previous year.
2.   We are considering a summer campaign, so we consider all customers who are defined as active in May the previous year (this could be earlier both for the modelling and for the application if a longer lead time into summer is required, e.g. with fashion clothes). Or we may be dealing with 'learning' or 'trialling' campaigns.
3.   We consider all members of the trialling campaign, for example, it could be past telephone activity or a general product testing campaign that has been sent out to a selection of customers (often chosen by the marketing department).

We distinguish our sampling frame population from the population of people on whom we apply the knowledge gained from the modelling. So the sampling frame for Christmas campaigns would be all the customers in October of the current year, and the sampling frame for the summer campaign would be all the customers in May of the current year, or for case 3 given earlier, the actual population when the modelling has been completed.

A typical example for this kind of problem is when you send out your main communication material, for example, your annual catalogue or the 'Big Book' for a mail-order house. In this case, you want to reach as many customers as possible on the one hand, but on the other hand, you want to keep the cost as low as possible.

So, for example, for the catalogue, the sampling frame would be all the customers who were sent the catalogue last time, or everybody in the records at a particular point in time, for example, June 1. It is important that the date corresponds with the time of year for which you wish to apply the models and that you will have enough time afterwards to measure the observed values of the variables.

Target variable: The target variable could be 'buying' or 'not buying' in response to the campaign, a binary variable. This is at the customer level in this example, so that each customer either buys or does not buy in a defined period, for example, in the first two months after sending out the catalogue. Alternatively, we may like to consider the whole period of time during which the catalogue is valid because we are optimising for the general reaction to the campaign. We recommend looking for every purchase that the customer made during the time period not only purchases that are clearly indicated as being out of the catalogue. Meta-data may also be available such as whether the customer buys a specific product or set of products, or the customer's general sales quantity and/or sales value at the chosen time. The chosen time is Christmas in 1 provided earlier or summer in 2 or for the trialling campaign in 3.

Note that we are basing the customer selection on whether they are more likely to buy than not. The specific purchase that the target relates to must therefore be highly relevant to the new promotion being planned.

Input data – must-haves: This includes raw data on historical purchasing behaviour, not just in response to previous campaigns, if any, but general historical behaviour. Raw data on historical reactions to previous marketing activities may be in the form of binary variables such as the customer recommended a friend to a publisher, or returned a questionnaire, or used a voucher. The raw data should include full details, for example, date, kind of product, amount, value, reductions, for example, buy one get one free, salesperson or channel of purchasing such as department store, online shop and local shop. Note that there will be different numbers of data items for each customer.

Raw data includes demographic and other data, for example, customer postcode, age, gender, longevity of customer, preferred channel, channel of first contact, source and family size. These details will not be static but will need updating as the customer channel preference or family size may change. If we are considering Business to Business (B2B), then we also need to consider company size, company sector (e.g. steel, oil, coal) and branch location.

Historical changes should be included at the customer level, for example, changing from one customer worth segment to another, moving address and changing marital status.

It is recommended to try to get the information for at least the last three years – five is even better. If your company is dealing with slow-moving products like cars and washing machines, then you need to go back for a much longer time period, depending on the expected cycle of product replacement or successful cross-selling actions.

Input data – nice to haves: This includes raw data on the historical details of advertisements, promotions and other communications sent to the customer; all complaints that are not related to the buying process itself, for example, asking for status of commitment in terms such as 'when can I stop my contract?'; details of online facilities used by the customer in login areas in your website, for example, looking for new, upcoming cars, books or other products or services (assuming they are featured on your website); whether customers recommend other people; and paying behaviour, for example, credit card, late payments, reminder letters, paying by cash, instalments or full amount.

Expected proportion of total time needed to prepare data: Preparation time is expected to be at least 70% of the total time needed to do the data mining, although some authors say 60–90%. It all depends on issues like the state of the data, the availability of algorithms and your skills as a programmer.

Data mining methods:

- Decision trees
- Regression, especially logistic regression
- Neural networks

How to do it:

Data preparation: The main issue is that after all the steps of transformation and aggregation, you should have one row for every customer but beware you can end up with thousands of variables.

Business issues: Before you start the next step, think about the potential implementation. Any step or manipulation that you know will not be possible to do in the final implementation should not be used in the learning phase.

We need to identify any known exceptional periods or variables, for example, exceptional anniversaries or country effects like recession. These can sometimes be dealt with by adding an indicator variable or, if the period is small, by excluding data pertaining to that time period. Or you can transform the data from the exceptional period to the average of activity in previous years. These sorts of business issues also arise post-modelling.

Transformation: The purpose of transformation is to stabilise the final models. For this, we need to replace missing values, exclude outliers (or ameliorate them), smooth the data, Normalise to use methods requiring Normality and reduce effects of changes in mean and variance. Note that in the decision tree method, the transformation can be done by the algorithm (usually grouping the variable into a small number, e.g. 10, of groups). Transformation is very important to improve results; otherwise, models can be over-affected by vagaries in the data.

Here, we consider suitable transformations of the input data, for example, we may have data on the number of books bought on topic A in the last 18 months by 250 000 customers. The values are ranging from 0 to 83. Instead of using these raw values, we can categorise them using a variety of methods including quantiles, Box–Cox and other analytical transformations. We recommend using the quantile method for this recipe. Especially for the catalogue example, it is very likely that after using the quantile method, the outcome for more rare variables is that they are binary. This helps to highlight the fact that being active is more important than the actual amount of purchases.

Check that the quantile transformation gives values that coincide with the unwritten rules of the business, for example, that customers from rural areas are grouped together, because, for example, ordered postcodes do not give a good grouping. In the case that single variables are filled with numbers representing codes, you have to do your grouping by hand or create dummy variables for the codes.

Marketing database: Detailed data from several sources needs to be combined into a marketing database with information for a single customer in one (long) row. The next part assumes that you have your data in this marketing database format (see Figure 8.2).

Analytics:

Partitioning the data: Note that patterns found in a representative sample of the data are usually a good indicator of patterns in the whole dataset. We develop our model on a learning sample, then check it out on a test sample and finally validate it on a further data subset. We take a stratified random sample based

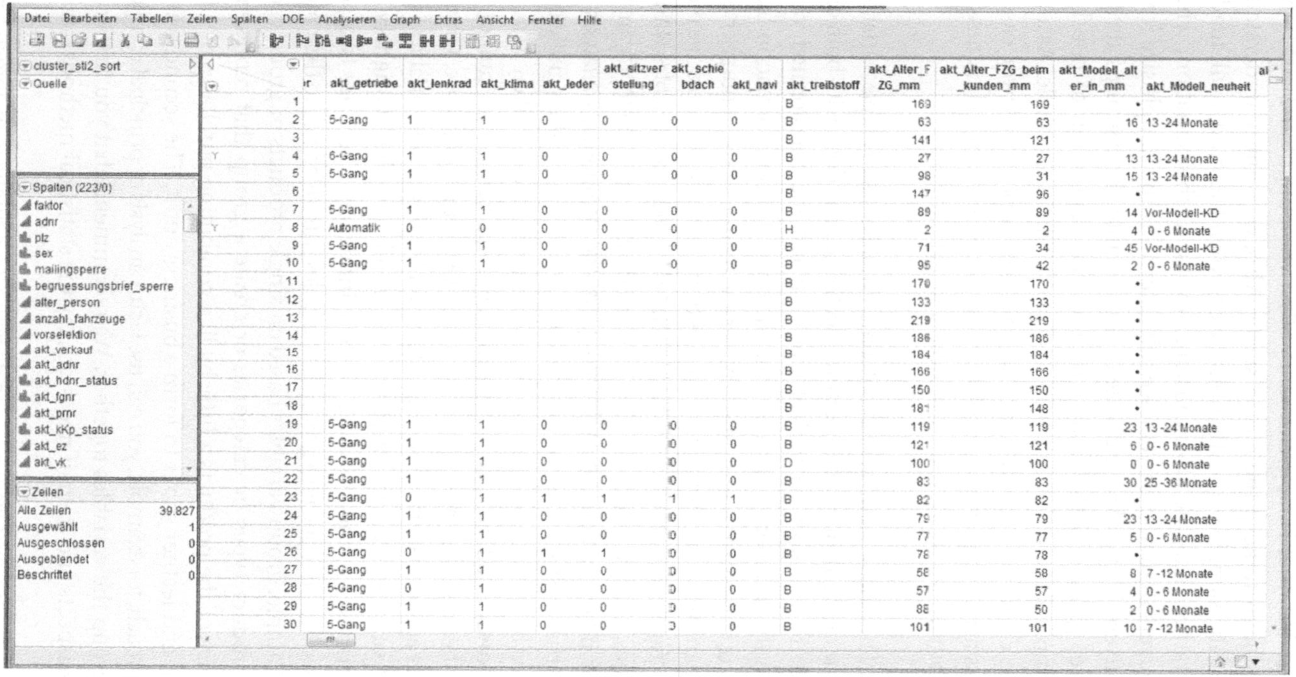

FIGURE 8.2 Example screenshot from a marketing database.

on the target variable to ensure that we have enough people in each category of the target variable, that is, approximately 50% who bought and 50% who did not buy. This helps to keep the model stable. Also, we want to understand the buyers, and if the sample is mostly non-buyers, then the model will tend to fit non-buyers rather than the all-important buyers. In this situation, when we have fewer buyers than non-buyers, we will determine the sample size from the number of buyers. Typically, we divide the buyers into three groups at random and take three groups of the same size from the population of non-buyers. Resulting sample sizes may range from 600 to 60 000, for example. As a rule of thumb, if the customers are very active, you can work with fewer cases, but if they are less active (e.g. buying only once or twice a year, or in business terms they do not buy very often), then you need a larger sample.

Pre-analytics: This enables us to screen out some variables, for example, variables that have zero value or are all one value. Feature selection can be done with both original variables and the transformed variables or with just the transformed variables depending on the number of variables and the degree of transformation that was required. Using chi-square testing and Cramer's V requires a small number of categories so they are best carried out with variables transformed by the quantile method. For variables transformed by the Box–Cox method, or original variables with many different values, feature selection can be carried out using methods including parametric or non-parametric correlations, principal components analysis, factor analysis or partial least squares. We continue the analysis using the significant variables identified by the feature selection. These input variables are likely to be highly correlated. However, all variables are potentially important so we cannot ignore variables just because they are highly correlated. Using the chi-square, Cramer's V tests and other appropriate methods helps to retain the important variables.

Regarding variables that have zero value or are all one value, it might be useful to group variables together to increase their influence. There are two general ways to do this. The first one is to group the variables with reference to their business aspect, for example, group variables based on single products or group variables relating to product types or categories or create a group that represents the common usage of the products a little bit like combining purchases of a formal shirt, tie, suit and leather shoes to form a business outfit. The second way is the statistical approach, for example, you can use factor analysis or principal components analysis to group the variables together. We recommend that you first consider the business aspects and then, if necessary, try the statistical methods later.

Model building: The preference is for decision tree analysis or logistic regression. In either case, the target is a binary variable with just two values such as 0/1, buy or not buy. The input variables do not need to be Normal for decision trees; Normality is more important for logistic regression. Our transformations help to make the input variables suitable. Note that the model is likely to need updating after the next stages of evaluation and validation. In some businesses, the model is referred to as a scorecard.

Evaluation and validation: Use the model on the test sample and other samples and compare observed and predicted buyers. Note the confusion matrix, or for better control, use a detailed report (or gain chart) that is free from cut-offs so that you can see how the score developed. Note also the sensitivity, ROC and lift charts.

Check whether the rules found fit with the business, for example, the variables in the model should coincide with 'gut feeling', and you should also feel at ease with the variables that have not been included. If there are variables in the model whose inclusion cannot be understood, this suggests that the model is over-fitted and something is wrong with it.

We also have to look for special effects which show up via the model which we were not aware of or had underestimated or forgotten. These could be specific to particular features that only occurred in one year such as big sales during the special 50th-year anniversary of the company or some other abnormal behaviour like halving the amount of advertisement in a past year. We should note any special aspects of the model and allow for them in the predictions. Finally, we check again that the model is useful for the business.

Implementation: Here, we address the original statement of the recipe, that is, how to name and address the right number of customers. We will choose the right number of customers based on the charts. The first filter might be permission-based restrictions or the availability of the planned contact channel. In the example, this restriction shrinks the population down to 639 238 as shown in Figure 8.3.

In this example, the criterion is based on a cut-off of lift = 2. All customers in groups with lift 2 and above will get the treatment. We need to consider how to implement the model on the actual data. Are there any special needs, for example, exceptions that cannot be transformed, or programming issues?

Note that the distribution of the customers in the sampling frame should be similar to that in the actual population, so that the learning is representative of the actual data.

The implementation can be done shortly after modelling, or it is also possible to implement a model regularly, for example, every month on new data or a long time later as in cases where you use a model just once a year. So you develop the model

	A	B	C	D	E	F	G	H	I	J	K
1	**descending score quality for class with 50.000**										
2	all customers						class 0 and class 1				
3	Score Class	Number Customers	Score P_ZIEL_AU_HYB1	Lift			Score Class	Number Cus	Score P_ZIEL_AL	Lift	
4	0	49999	0,866	2,36			0	9999	0,9181	2,51	
5	1	50000	0,740	2,02			1	10000	0,9169	2,50	
6	2	50000	0,640	1,75			2	10000	0,8947	2,44	
7	3	50000	0,571	1,56			3	10000	0,8112	2,21	
8	4	50000	0,418	1,14			4	10000	0,7913	2,16	
9	5	50000	0,326	0,89			5	10000	0,7759	2,12	
10	6	50000	0,281	0,77			6	10000	0,7733	2,11	
11	7	50000	0,274	0,75			7	10000	0,7372	2,01	
12	8	50000	0,191	0,52			8	10000	0,7073	1,93	
13	9	50000	0,132	0,36			9	10000	0,7073	1,93	
14	10	50000	0,128	0,35			rel sum	99999	0,80332	2,19	
15	11	50000	0,099	0,27							
16	12	36239	0,031	0,08							
17	sum	636238	0,366	1,00							
18											

FIGURE 8.3 Example screenshot of output.

once and then you might use it for a longer time. As a rule of thumb, as the model becomes 'older', it may not fit so well. Whether 'older' means a couple of months or five years depends on your actual business. This is an important point, and it is sensible to identify suitable measurements to monitor the quality of the model over time. One possibility is having control groups with no special marketing and comparing them with a group subject to marketing activities. The comparison measure is the relative lift. Another method is periodically to compare the ranking as regards the reaction of a sample of customers after marketing activities.

In general, you have to follow these steps to implement the model:

- Carry out relevant aggregation and transformation for all variables that are part of the model to determine if there are any special needs, for example, exceptions that cannot be transformed, or any programming issues.
- Compare the sample distribution of actual (new/current) aggregated and transformed variables to the distribution they had in the population to check they are similar. If the distributions are similar enough, you can go ahead, but if you notice that there are big differences, you have to clarify the reasons for the differences and how they may affect the results. If necessary, you have to go back and do some new modelling on a learning sample that corresponds better with the actual population.
- Try to use the model to estimate the affinity to an offer. Depending on the data mining method you choose for modelling, it may be that you can use your data mining software to do the estimation. However, it may be that this is not feasible because the software is too slow or the system where you plan to use the scores/estimators cannot communicate with your data mining software. If the model estimation is likely to encounter problems, then you may have to transform the model to a group of rules that can run on other programming languages.
- Prove that the distribution of the estimates corresponds to the distribution on the learning sample.

Note that the distribution of the customers in the sampling frame should be similar to that in the actual population, so that the learning is representative of the actual data.

Hints and tips: If you have a lot of variables, replace the original variables by the transformed variables in the data file to make the data file as small as possible. The downside of this procedure is that you then have two versions of the file, one with original variables and one with transformed variables.

In difficult cases where you are unsure of the model quality or it does not fit with the unwritten rules of the business or it seems that the model fits the sample but when generalised or applied to the test and validation data sub-sets, it is apparent that it is over-fitting, then try model building using several samples and compare the results in each sample, taking the variables forward that occur most often. You can also gradually build on your experience of dealing with similar situations or customers and let this influence your final choice of model. Note that the important issue is whether the model is validated and if it is useful when implemented.

Nobody will care later on that it was a great model in the learning phase; they only care about the quality and separation ability of the implemented model. If possible, make sure that you can run a small example test to demonstrate and prove the quality of your models.

How to sell to management: For management, the model is good if it saves money regardless of how good it is in its statistical aspects. Show management money instead of statistics and numbers. In this case, show them that they waste money if they communicate with those who are unlikely to react to this offer. The easiest way is to show it with numbers, such as ROI.

Determine the % of customers to be targeted from the model on the learning set. Apply this % to the population. Predict sales response from the sample of target customers found from modelling on the learning dataset and estimate the value of their purchases. Additionally, we can use the real values of the sales from the sampling frame population.

We can compare this with the sales lost to the people not now targeted and compare the saving in effort to the previous return.

Compare the number of customers to be addressed with the population base and calculate the saving in actual money currency terms. For example, if each contact costs £0.50 and the population is 500 000 and we have reduced it to 60 000 relevant customers for this offer, then the savings are (500 000 – 60 000) × 0.50, which is £220 000 saved.

8.2   RECIPE 2: TO FIND THE *x*% OF CUSTOMERS WITH THE HIGHEST AFFINITY TO AN OFFER

Recipe 1 was concerned with response optimisation and finding the right number of customers to contact in a campaign. The recipe considered here differs from Recipe 1 in that we wish to find a fixed number or % of the population with the greatest affinity to an offer rather than the 'right' number of customers.

Finding customers with the highest affinity to buy an offer is just another way of implementing the model. In Recipe 1, we are looking for significant jumps in the detailed lift or gain chart that indicate a change in the level of affinity; the assumption in Recipe 1 is that such a jump indicates a natural border of affinities from which we can extract the right number of customers to contact (Figure 8.4).

In this recipe, the business question starts with a given number or percentage of customers, and we need to find the best customers up to the required number. For example, assume that the marketing budget for a given campaign is fixed at 25 000 euro and the overall cost to market to a customer in that campaign is 1 euro/customer. The maximum number of customers to be selected for that campaign is 25 000 customers. If the total number of customers is bigger than 25 000, for example, 80 000, then we need to select the 25 000 best customers to whom to promote the offer in that campaign. If there are fewer than 25 000 customers in total, then they could all be selected. Alternatively, after a discussion with your marketing colleagues, it might be decided to reduce the number of customers for the campaign in a similar way to Recipe 1. In this case, the lift chart would be used to make the decision on how many customers are relevant for the campaign and not all the money would be spent.

Target variables: The target variable will be designed according to the main offer of the campaign. If it is a price-sensitive campaign, then the target might be created by all customers who bought a reduced price product or a special offer in a certain time. In any case, the target will be a binary variable.

Implementation: In general, the implementation is similar to Recipe 1. The model determined by the data mining is used to calculate the affinity to the offer for each customer in terms of their predicted target score. The customers are then placed in the order of descending score, and the top 25 000 best customers are selected.

Sometimes, it may happen that the required % of customers falls across two groups. For example, if we want the top 10%, we would have a problem if including just the top two groups gives 8% but including the third group gives the top 15%. This problem is avoided in decision tree analysis by allowing a further branch split or equivalently doing less pruning.

8.3 RECIPE 3: TO FIND THE RIGHT NUMBER OF CUSTOMERS TO IGNORE

This recipe varies from Recipe 1 in that we now want to identify customers who are least likely to buy (see Figure 8.5).

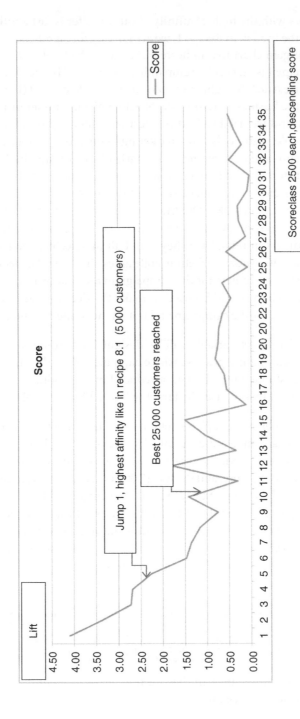

FIGURE 8.4 Lift chart to detect potential cut-offs.

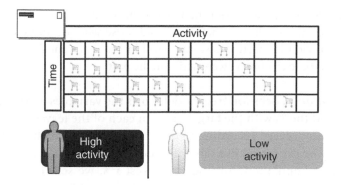

FIGURE 8.5   Activity pattern.

Target variables: The target is normally a straightforward binary variable. Even when looking for the worst customers, it is not recommended to reverse the target – accepting an offer should still be marked with '1' in the target. From the statistical point of view, it makes no difference to the modelling whether accepting is scored as 1 or not accepting is scored as 1. However, as far as communications are concerned, it might be difficult to tell your colleagues that those customers with the lowest affinity for an offer will have the highest scores. It is much easier to say that those with the lowest affinity also have the lowest scores.

In our practical problem, we want to base our decision on those customers who are not very likely to buy at all. Using domain knowledge we might, for example, note a rule of thumb about the relationship between the number of days people did not buy and the likelihood to buy again. As we are concentrating on customers with low expected affinity to buy, it is highly likely that they had no or very few activities in the last couple of months. This indicates that it might help to get better results in prediction if we split the population into groups with different activity patterns. For example, we may have one group for everybody with some activity during the last 12 months and another group for those with no activity in that time. If you do not split the group, there is a risk that the data patterns of the active customers are so strong that they dominate the models and you cannot make real decisions on the less active customers who are the ones for whom you need the decision.

If you split the data, you have to build at least two models, one for the active group and one for the inactive group. It is possible that you have to do two kinds of modelling, which will result in two very different models. It is quite likely that the model for the more active customers fits much better and has a totally different variable structure. The model for the inactive customers is quite likely to be poorer than the model for the active customers or any other model built from the whole sample. However, separating out the inactive customers will help you to distinguish between those who are 'half-dead' and those who are 'dead' because the model for previously inactive customers will include relevant data for the time period when these customers were active in the past.

Implementation: A decision tree model shows the variables used to form the branches and the % with the target value in each of the leaves. The decision rules are then ordered according to the % with the target value. The results can usefully be expressed in a table. The decision tree model is then applied to the whole population. To find the lowest x%, we select the customers corresponding to the decision rules with the lowest % target values until we obtain an optimum number such that adding any more will give an unnecessarily large number of customers for a small change in affinity, as in Recipe 1.

8.4   RECIPE 4: TO FIND THE X% OF CUSTOMERS WITH THE LOWEST AFFINITY TO AN OFFER

This recipe is very similar to Recipe 3 and is the negative twin to Recipe 2. Instead of finding the best, you are looking for the worst customers to give them a different marketing treatment or to exclude them completely. In Recipe 3, the number of customers to be excluded is indicated by statistical

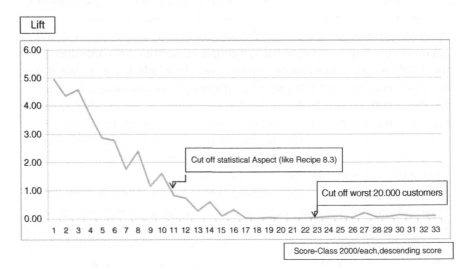

FIGURE 8.6   Lift chart showing cut-offs according to different criteria.

aspects of the model; in this recipe, the number of customers is dictated by marketing considerations related to the existing reporting and forecasting systems. The decision is influenced by ROI aspects and the overall communication strategies.

It is possible that you have to combine the aspects of Recipes 2 and 4 for one marketing campaign, for example, for communication reasons, the top 20% of all customers will get a mailing, the middle 40% will get an email, and the worst 40% will get no communication at all this time (see Figure 8.6).

Implementation: In general, the implementation is similar to Recipe 3. The model determined by the data mining is used to calculate the affinity to the offer for each customer in terms of their predicted target score. The customers are then placed in the order of descending score, and the bottom 20 000 worst customers are selected.

8.5 RECIPE 5: To FIND THE *x*% OF CUSTOMERS WITH THE HIGHEST AFFINITY TO BUY

This recipe just considers the general affinity to buy anything in a defined time slot. The difference between Recipe 5 and Recipe 2 is the target. Recipe 2 is specific to a special offer or a particular purchase. In this recipe, the target variable is any purchase of any kind, or sometimes, it may just be any activity or any response at all (see Figure 8.7).

The kind of predictive modelling in Recipe 5 is often used if you want to qualify people for a VIP programme or special services, such as a dedicated telephone number or login to the website; or to send out invitations or brochures

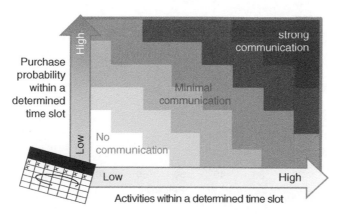

FIGURE 8.7 Communication strategy related to general buying behaviour.

or flyers that include different types of products or, for example, an occasional catalogue. Another application is when you want to use a general approach to predict those prospects with the highest overall affinity so that you are likely to turn them into customers.

Target variables: The target is just the general buying behaviour; as long as you purchased at least once in a given time period, you will get the Target = 1 value; otherwise, if you have no buying activity, you get the Target = 0.

Implementation: Implementation is similar to Recipe 2.

8.6   Recipe 6: To Find the *x*% of Customers with the Lowest Affinity to Buy

This recipe considers the general affinity to buy anything in a defined time slot with the aim of selecting people for a reactivation communication. The difference between Recipe 6 and Recipe 4 is that Recipe 4 is specific to a special offer and Recipe 6 is based on general affinity.

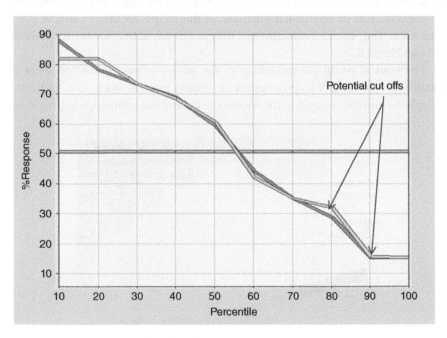

Figure 8.8   Lift chart to help decide who to try to reactivate.

It is important to cross-check that the relevant input variables are meaningful for those inactive customers that should be reactivated. For example, if a customer did not buy anything during the last 12 months, a variable that describes the buying behaviour 6 months ago is irrelevant to support the decision on those inactive customers who last bought 12 months or more ago. But a variable that describes the buying behaviour 18 months ago might help to discriminate the affinity to buy/ to be reactivated of those with at least 12-month inactivity (see Figure 8.8).

Target variables: The target variable is just the general buying behaviour; as long as you purchased at least once in a given time period, you will get the Target = 1 value; otherwise, if you have no recent buying activity, you get the Target = 0.

Implementation: The implementation is similar to 8.4.

8.7   RECIPE 7: TO FIND THE *X*% OF CUSTOMERS WITH THE HIGHEST AFFINITY TO A SINGLE PURCHASE

The business question behind this recipe is to find those customers who are not likely to buy more than once in a time period, or if you offer a long-term contract as well as single purchases, this recipe aims to find those people who will only go for the single purchase. Another variation is to find the customers who will just buy a single product regardless of whether a single piece or a number of pieces of this article is offered in the given period. This recipe is just a variation of Recipe 2 with a different target; defining the correct target is the main issue. This is illustrated in Figure 8.9.

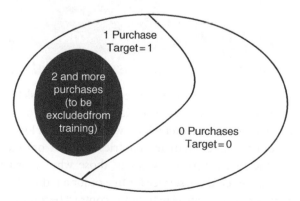

FIGURE 8.9   Purchase behaviour to be included in the recipe.

Example	Target=1 People Are Those with:	Target=0 People Are Those with:	People to Disregard Are Those with:
Single order	One distinct order date (one or more products may be involved in order)	No order at all	Two and more distinct orders
Single purchase	One distinct order date (One or more products involved in order) and part or full payment of that order (no 100% returns)	No order at all	Two and more distinct orders. One distinct order date and no payment
Single article or product	One distinct order date with only one item of a specific product	No orders at all	Every other order

Figure 8.10 Typical definitions of target variable.

Target variables: The target has to be chosen very carefully, and you should clearly define the Target = 1 and the Target = 0. Your Target = 1 are people with a specific buying behaviour that can be described exactly. Note that there are other people around who also made some purchases but who might not fit your target definition. In other words, it is very likely that Target = 0 is not the opposite of Target = 1. There might be a major group of customers who are neither Target = 0 nor Target = 1.

Typical definitions of targets in a given time period are given in Figure 8.10.

8.8 RECIPE 8: TO FIND THE *x*% OF CUSTOMERS WITH THE HIGHEST AFFINITY TO SIGN A LONG-TERM CONTRACT IN COMMUNICATION AREAS

Typical examples where this recipe may be needed are for newsletter or magazine subscriptions, telephone contracts, web services, sports clubs or other memberships. In this case, you wish to predict who is willing to sign a long-term contract for a specific service or product. In many countries, 'step out' or cancelling options are given during the first couple of weeks to those who signed the contract. Sometimes, people are offered the service for free or for a reduced fee so that they can try it out, but if they do not cancel in time, the contract becomes valid automatically (see Figure 8.11).

FIGURE 8.11 Get a smartphone for less if you sign in.

Challenge: The long-term contract poses a different situation to making a purchase online, in a department store or from a mail-order house where a new data entry is created each time a purchase is made. These long-term contracts produce smaller datasets, and you have less interaction between customer and sales organisation because the customer will acquire and pay for their product or service in a fully automated way with no interaction on the customer's side. The process itself depends very much on national habits. But the specific characteristics are always the same.

Normal/single purchase process: After the first purchase, the customer decides actively to re-buy at a time that suits them. If they stay passive, then the process stops and no purchase is made.

Long-term contract purchase process: After the first purchase/signing the contract, the customer can stay passive, and the product or service just continues to arrive and be available. In other words, if the customer stays passive, the process continues.

Data: The potential sources of data are quite similar to those in the other recipes. However, in comparison to other recipes, the type of input variables is different, and you have to focus much more on any activity of the customers because there are fewer customers registered in the system. If available, data on service usage can be used as input variables as well, for example, how often has the software been used during the last three weeks? It may also be possible to notice changes in the usage that can be transferred into input variables, such as the customer may significantly increase the amount of overseas calls in the last four weeks compared with the observed pattern in the last year.

Pre-analytics: Possible data transformations need to be considered, but in any case, it is very important to use and transport domain knowledge into the input variables.

Target variables: The target variable depends on the given business model; most of the time, it might be better to count those as Target = 1 who are in the situation that 'early day' cancelling is not possible any more. It is important to differentiate between those who are still more or less in a 'trying out' mode and those who have signed up and are locked in with specific cancelling options like having to give three months' notice. So if you focus on those with fixed signed contracts, just use them as Target = 1 and reject those with other more or less freer options.

Modelling, validation and implementation are straightforward as described in other recipes; you can choose between any kind of regression, decision trees and neural networks.

8.9   RECIPE 9: TO FIND THE X% OF CUSTOMERS WITH THE HIGHEST AFFINITY TO SIGN A LONG-TERM CONTRACT IN INSURANCE AREAS

This case is a variation of Recipe 8. The key issues are dominated by domain knowledge. Insurance habits and culture depend on local habits and circumstances as well as on legal requirements and the likelihood of individual risk (see Figure 8.12). In the insurance industry, the offers can be separated into three categories. These are listed in the following with some examples of typical situations found in the German market:

- Insurance that is required by law: health insurance and car insurance
- Insurance that is recommended as a 'must-have': personal liability, occupational disability insurance and household insurance

FIGURE 8.12   Features of insurance.

- Insurance that is taken to cover personal or local risks: life insurance and legal protection insurance

Especially for insurance required by law and insurance that is recommended as a must-have, it may make a bigger difference whether your potential customer is just contracting their first own insurance (first ever or first ever with your company) or whether they are already in an insurance contract but are willing to churn and to accept a better offer given by the same company or a competitor.

Those who are willing to make their first contract with the company have to be modelled separately in a similar way to that described in Recipe 1. The main difference between those and the customers who are already known is the accessibility of detailed historical data. To develop a model for those more or less new customers, you can only use publicly available data as enrichment data and data given by the customers themselves. The other older type of customers can be handled as in a more general example of predictive modelling such as Recipe 8.

Target variables: It depends a bit on the contract itself, but the target can be counted as 1 as soon as any reassignment times are over. Target = 0 are all those who did not react to the offer at all. Those people who respond but step back from the contract at resignation time are not part of the data used for any kind of modelling.

Implementation: These kinds of models are used to optimise sales and marketing campaigns, so the results of the implementation and the model itself will be stored in the campaign management tool. In some companies, the predicted affinity to buy is also stored in the service desk/operational CRM tool to provide the employees with the necessary information on single product affinities to encourage cross- or up-selling activities.

9

Intra-Customer Analysis

The following recipes describe how to make the best of customer information whether in a data mart fed from a data warehouse or from a more *ad hoc* data source. Descriptive analytics can derive rich information. We consider the

A Practical Guide to Data Mining for Business and Industry, First Edition.
Andrea Ahlemeyer-Stubbe and Shirley Coleman.
© 2014 John Wiley & Sons, Ltd. Published 2014 by John Wiley & Sons, Ltd.
Companion website: www.wiley.com/go/data_mining

use of cluster analysis and association rules as well as decision trees, logistic regression and Self-Organising Maps (SOMs).

9.1 RECIPE 10: TO FIND THE OPTIMAL AMOUNT OF SINGLE COMMUNICATION TO ACTIVATE ONE CUSTOMER

This is a descriptive analytics problem. The data collected by the client will be used to find hidden relationships between the variables and the buying behaviour of the client. It is not so necessary to have fine detail in the variables that are likely to be of major importance when we are just looking at the general relationship between the variables and the buying behaviour. Therefore, the variables have been put through the binning process, and the descriptive analysis proceeds with the binned values. The work in this chapter also relates to Recipe 1 (see Figure 9.1).

Input data – must-haves: The recipe requires the full communication history and the buying history.

Target variables: The following could be useful as target variables: number and quantity of purchases or being active or not.

Data mining methods: Suitable descriptive methods include:

- Histograms and multiple bar charts
- Contingency tables and chi-square analysis
- Scatterplots and correlation

We could plot the response for different numbers and types of communication.

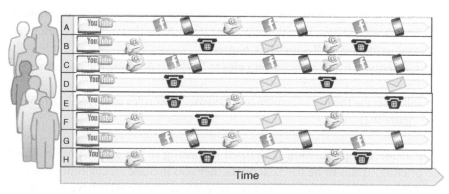

FIGURE 9.1   Activating customers.

The binning and the analysis of relationships between variables and buying behaviour help to clarify how to find the optimal amount of single communication to activate one customer.

9.2   RECIPE 11: TO FIND THE OPTIMAL COMMUNICATION MIX TO ACTIVATE ONE CUSTOMER

Industry: The recipe is relevant to everybody using direct communication to improve business, for example, mail-order businesses, publishers, online shops, department stores or supermarkets (with loyalty cards).

Areas of interest: The recipe is relevant to marketing, sales and online promotions.

Challenge: The challenge is to find the right combination of direct communications to activate a customer. Initially, as always, we will have a business briefing to determine exactly what is required.

In marketing, you have the problem that you are never totally sure what kind of campaign or combination of campaigns is successful. If you talk to experienced marketing colleagues, they will tell you that you need to build up awareness by communication to get reactions. This means that it is not only the last communication that is responsible for the reaction, but it is the combination of all previous communications. For example, a clothing manufacturer may have carried out the following combination of marketing activities over the last half-year: vouchers sent to selected customers, recommend a friend, buy one get one free and then finally they will get a reaction.

There are different ways to determine the optimal combination of communications:

1. Describe the problem so that it can be treated as a prediction problem. The challenge here is to describe the target. After that, you can work in an analogous way to Recipes in Chapter 8.
2. Use sequence and association rules to find and describe the patterns.

We will describe the use of sequence and association rules to solve this problem (see Figure 9.2).

Population: The population is defined according to the actual problem and the briefing. It has to be discussed whether you include all customers (active and inactive) or just those who are currently active or who have only been inactive for a short while in the past. Note that inactive customers will still

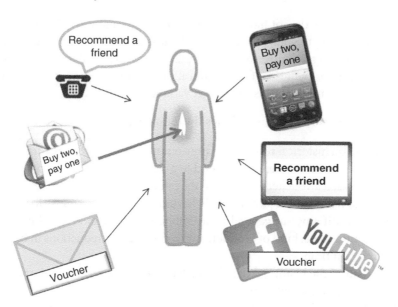

FIGURE 9.2 Communications sent to activate a customer.

have received some kind of advertisement and that some of them may become active later on.

Depending on the number of customers, it might be useful to carry out sampling. If you decide to sample, the sampling has to be carried out on the customers and not on the records of advertisements or other promotional activities. You need to get a sample of customers and add all relevant data to them. Note that the distribution of the customers in the sample should be similar to that in the actual population so that the learning is representative of the actual data. The representativeness of the sample should be checked, and if necessary, some changes should be made. For example, a training sample was found to have more people in the working class than appear in the actual data. The sample was modified by taking a better stratified sample, and the model was rebuilt. Note that the population of interest may well be a subset of the whole customer database. We may also decide to further sub-divide if we think that different sequences and associations are likely with different groupings.

Necessary data: The areas of data that could be used must contain information on all kinds of advertisements and campaigns that could reach a customer and on the general customer reaction not only their direct response to one advert or communication.

Target variable: We do not have a target in the traditional way. The objective is more about trying to control and learn. The instructions will differ depending on the data mining software you use, for example, with SAS Enterprise Miner if we are looking for patterns of communication, then we will need to nominate the type of communication as the "target variable".

Input data – must-have: This includes the following:

- Raw data on historical reactions to previous marketing activities, for example, the customer recommended a friend to a publisher, or returned a question-naire, or used a voucher, in between the relevant time.
- Raw data on historical purchasing behaviour not just in response to previous campaigns, if any, but general historical behaviour. We include full details, for example, date, kind of product, amount, value and reductions in between the relevant time.
- Raw data of the historical details of advertisements, promotions and other communications sent to the customer.
- Raw data on all complaints that are not related to the buying process itself, for example, asking for status of commitment in terms such as 'when can I stop my contract?'.

Example: The first few lines of typical data showing the input data (to be extended in the transformation section) are shown in Figure 9.3.

Data mining methods: The following methods can be used:

- Sequence analysis
- Association rules

How to do it:

Data preparation: The main issue is that after all the steps of transformation and aggregation have been done, you should have several (at least three) rows for every case (customer). Therefore, the data is in customer order with multiple rows for each customer.

Business issues: Before you start the next steps, think about the potential implementation. Any step/manipulation that you know will not be possible to do under implementation should not be used in the learning phase. It is very likely that the results will be used to guide future marketing campaigns automatically, so note that you must be able to implement the rules in your campaign management system.

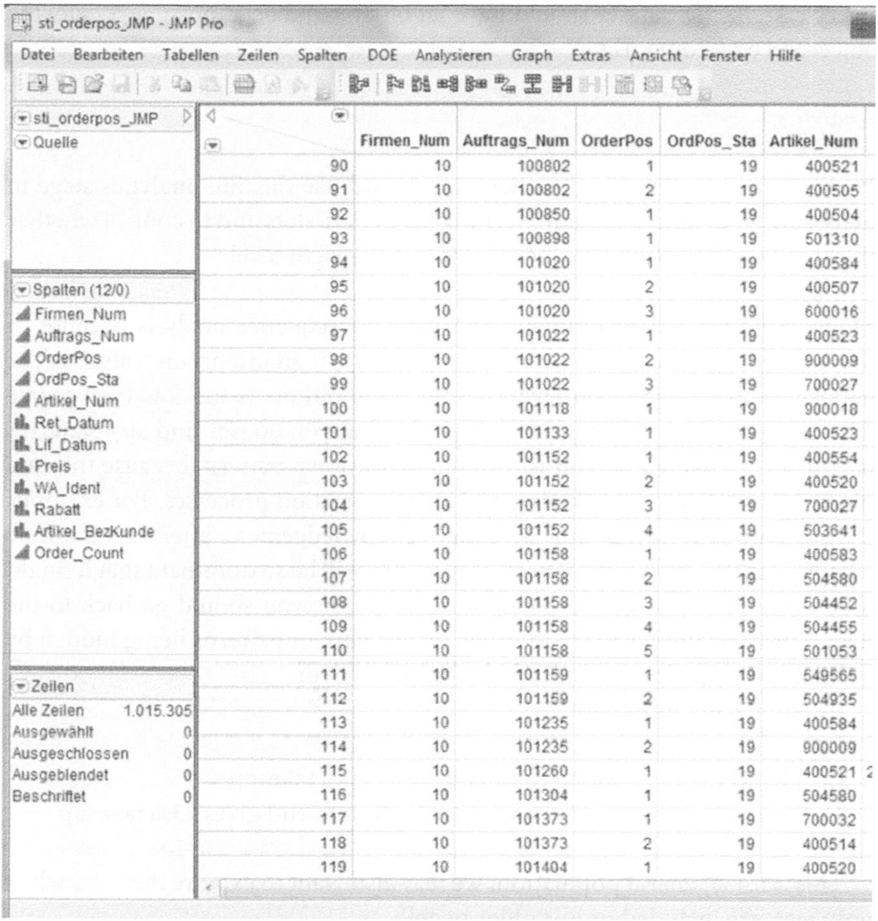

	Firmen_Num	Auftrags_Num	OrderPos	OrdPos_Sta	Artikel_Num
90	10	100802	1	19	400521
91	10	100802	2	19	400505
92	10	100850	1	19	400504
93	10	100898	1	19	501310
94	10	101020	1	19	400584
95	10	101020	2	19	400507
96	10	101020	3	19	600016
97	10	101022	1	19	400523
98	10	101022	2	19	900009
99	10	101022	3	19	700027
100	10	101118	1	19	900018
101	10	101133	1	19	400523
102	10	101152	1	19	400554
103	10	101152	2	19	400520
104	10	101152	3	19	700027
105	10	101152	4	19	503641
106	10	101158	1	19	400583
107	10	101158	2	19	504580
108	10	101158	3	19	504452
109	10	101158	4	19	504455
110	10	101158	5	19	501053
111	10	101159	1	19	549565
112	10	101159	2	19	504935
113	10	101235	1	19	400584
114	10	101235	2	19	900009
115	10	101260	1	19	400521
116	10	101304	1	19	504580
117	10	101373	1	19	700032
118	10	101373	2	19	400514
119	10	101404	1	19	400520

FIGURE 9.3 Typical data about communications.

Transformations: Based on the fact that things, that did not happen, are not stored in the database, you have to think how to create dummy cases that represent this 'Nothing happened in a special time slot' issue. 'Nothing happened' has two sides: no marketing and no customer reaction. The length of an average time slot in month, weeks, days or hours depends on your business and has to be discovered in the pre-analytics phase. If you, for example, detect that a week as a time slot will suit your business, you have to create a dummy variable for every week without advertisement or campaign and for every week without a customer reaction. Advertisements have to be classified into sub-groups such as standalone emailing, telephone marketing, direct mail, catalogue and salesman. Customer reaction has to be classified as well, for example, order, complaint, unsubscribe and return.

Every case in the dataset should have an order number, if possible a date, the object and a customer ID.

Analytics:

Pre-analytics: As described earlier, you should use this pre-analytics stage to detect the right length of your 'nothing happened' slots and to control whether the chosen aggregation of your objects is at the right level.

Model building: We recommend the use of a sequence analysis because in the marketing context, the given order of objects is an important feature of the pattern. If the aggregation level of your observations is too low (too coarse with too many issues combined), it may be that you do not find any patterns. It is also very likely that you will find a well-known pattern, because they are the results of defined marketing and communication processes. For example, if your company sends out an email acknowledgement after every online customer interaction, this pattern of response will be so dominant that it might hide other really unknown patterns. In this case, you should go back to the data transformation and try to solve the problem of patterns being hidden by bundling such process items together into one item.

Example rules found are:

Email and then telephone campaign gives 20% take-up

Mail campaign, then email and then invite-a-friend gives 23% take-up

This is for a general population; we may also want to explore the rules when only men are included or just older people.

Evaluation and validation: Use the set of rules found on the test and validation samples and compare the observed results with the expected ones. We might compare the rules via the support and the confidence as described in the methods section.

An important question is whether the rules found fit with the business. If there are rules and patterns in the model that we cannot understand, and neither you nor anyone else knows why they are there, this suggests that the model is over-fitted and something is wrong with it.

We also have to look for special effects which show up via the model which we were not aware of or we had underestimated or forgotten. These could be specific to particular features, for example, something which only occurred in one year, or an abnormal reduction in the amount of advertising in a past

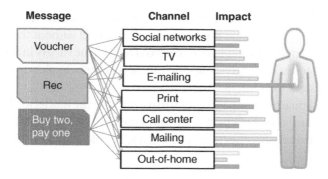

FIGURE 9.4 Communication mix to activate a customer.

year. Is the model useful for the business? Usefulness is judged by use and continued use and outcome in terms of financial benefit (see Figure 9.4).

Implementation: The implementation can be done shortly after learning, or it is also possible to implement rules regularly like every week on new data to run fully automated campaigns.

It is necessary to do every single aggregation and transformation for all variables that are part of the rules.

Hints and tips: If possible, make sure that you can show and prove the quality of your model by comparing the results obtained with and without the model. For example, it is a good idea not to use your model to select 100% of the people who will be contacted (or get the marketing treatment) but to allow some of them to be selected in other ways which could be the original method used before.

You can then keep a check on the performance of your model and show its superiority by comparing the average responses. If you are confident in your model, you could let as many as 90% of the customers be chosen by the model and 10% be chosen in other ways. Otherwise, in the early days of adopting modelling, you could select 50:50. Note that another way of allowing this comparison is to split the population in the required ratio and make the selection from each part.

There are other good reasons for making only part of the selection by the model. One of these is to ensure that you do not lose the opportunity to observe the reaction of a group of people who are not considered important in the model but do exist in the population. Making part of the selection completely at random ensures that you have a sub-sample whose responses you can observe. For example, if the model indicates that gender is important, then you could end up with only females being selected, and in the future, no one would know how men react. Also, there would be no way to illustrate the success of your model without the reaction of some men to compare it with. Finally, you need to keep the variety in the customer selection so that when models are updated, there is a chance to pick up features that may have changed their importance.

Note that sometimes management will want to use the model for 100% of the selection because they believe it is the most cost-effective method, and unless you can persuade them otherwise, you will not be able to make the comparison discussed earlier.

How to sell to management: From the point of view of management, the rules are good if they

- Save money, or
- Improve sales, or
- Increase customer loyalty

regardless of how good the rules are in a statistical sense. Show the management money instead of statistics and numbers. If you cannot prove that at least one of the benefits in the three bullet points applies, then your rules might not be used. In any case, show them that if the campaigns do not follow your rules, they will waste money because they will need more marketing budget and more campaigns to be done more often.

9.3 RECIPE 12: TO FIND AND DESCRIBE HOMOGENEOUS GROUPS OF PRODUCTS

Industry: The recipe is relevant to everybody using data to learn and improve their product categories.

Areas of interest: The recipe is relevant to purchasing department, marketing, sales and online promotions.

Challenge: The challenge is to find the right combination of products to improve your category management. For example, this is important for department stores. There are two alternative ways to view the problem:

1. The problem is about what is bought together – this kind of problem is solved by using association rules.
2. The problem is about which products have similar usage and customers (a problem that will be solved by using clustering methods).

We will focus on the second option (see Figure 9.5).

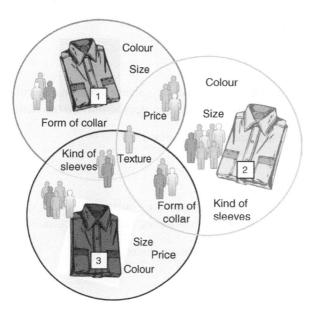

FIGURE 9.5   Typical application of cluster analysis.

Necessary data: The data must contain detailed information on buying habits and customer demographics. Information on advertisements is nice to have but is not that necessary.

Population: Note that in this recipe the products are the cases rather than the customers, so it is the products that make up the population. If you have no other information, it might be useful to define all products that are offered in one year as the population.

You have to decide on what level of detail you want to envisage a product as a product. For example, consider the following situation:

Your company sells shirts and it is up to the product manager to decide on which level the product is seen as a product. Note that the customer may have a different view to the product manager.

There are different ways to describe shirts within the data; the following are a few options to help you visualise the situation:

Option 1

Shirt, white, button-down-collar, long sleeves, buttons/no cuffs, size 17: product ID 140032 amount 1

Shirt, white, button-down-collar, long sleeves, buttons/no cuffs, size 15: product ID 140029 amount 1

Shirt, blue, button-down-collar, long sleeves, buttons/no cuffs, size 17: product ID 140056 amount 1

In option 1, there are three different products.

Or

Option 2
   Shirt, white, button-down-collar, long sleeves, buttons/no cuffs, amount 2
   Shirt, blue, button-down-collar, long sleeves, buttons/no cuffs, amount 1

In option 2, there are two different products.

Or

Option 3
   Shirt, button-down-collar, long sleeves, buttons/no cuffs, size 17, amount 2
   Shirt, button-down-collar, long sleeves, buttons/no cuffs, size 15, amount 1

In option 3, there are two different products.

Or

Option 4
   Shirt, button-down-collar, long sleeves, buttons/no cuffs, amount 3

In option 4, there is just one product.

   You have to use your domain knowledge to decide which level will help to
   solve the problem. The nature and significance of the potential results
   are influenced by it.

Target variable: There is no target as we are dealing with a method of unsupervised learning.

Input data – must-have: The must-have data includes:

• Raw data on historical reactions to previous marketing activities, for example,
  the customer recommended a friend to an outlet, or returned a questionnaire,
  or used a voucher, within the relevant time.
• Raw data on historical purchasing behaviour not just to the previous
  campaigns, if any, but general historical behaviour. We include full details,
  for example, date, kind of product, amount, value and discounts within
  the relevant time.
• Raw data of the historical details of advertisements, promotions and other
  communications sent to the customer.
• Raw data on all complaints that are not related to the buying process itself, for
  example, asking for status of commitment in terms such as 'when can I stop
  my contract?'

Data mining methods:

- Cluster analysis
- SOMs

How to do it:

Data preparation: In this example, the products are the cases. So when you start generating the data, keep in mind that you have to build up everything from the viewpoint of the product. But generally, this is not a problem, and you can use the same ideas as would usually be used when dealing with customers but make sure it is from the product viewpoint.

Business issues: Before you start the next step, think of the potential implementation. Any step or manipulation that you know cannot be done in implementation should not be used in the learning phase. It is very likely that the result will be used for decisions and strategic planning, rather than for immediate practice, so the technical considerations of the implementation might be minor.

Whether you work on the level of a detailed product number or on an aggregated level depends on your actual briefing. If in doubt, it is better to be more detailed. But the detail level must be viewed from the customer usage viewpoint and not from the supply chain viewpoint, where two products that are absolutely similar may get two product numbers because they are produced by different companies.

Transformations: As in other modelling questions (e.g. in Chapter 8), the same kind of transformations and aggregations should be done, but the cases are products so you have to modify it a little bit.

Analytics:

Pre-analytics: You should use this phase to control whether the chosen aggregation of your objects is on the right level and to get a descriptive overview of the special aspects and descriptions of the products. Depending on the number of variables created, this phase can be used to detect those variables that might have some impact on the clustering.

Model building: We recommend using a hierarchical cluster method. The dendrogram might help you to explain the results more easily.

Evaluation and validation: Use the set of cluster descriptions found from the training samples and compare them with those found from the test samples.

Do the rules found fit with the business? If there are rules and patterns in the model that we cannot understand, and neither you nor anyone else knows why they are there, this suggests that the model is over-fitted and something is wrong with it.

We also have to look for special effects which show up via the model which we were not aware of or we had underestimated or forgotten. These could be specific to particular features, for example, which only occurred in one year, for example, like a special 'birthday offer'. Sometimes, it is better to delete these products from the dataset and to redo the analysis.

Finally, you have to check whether the model is useful for the business.

Implementation: The results are mostly used for strategic decisions, so there are no technical barriers to be aware of.

Hints and tips: If you get the impression that the results seem to be a bit boring or peculiar, try a different kind of aggregation level for the product. Make sure that product numbers are unique or a historical dimension table for additional product information is available. Make sure that, for example, socks and washing machines do not share the same product number: for example, until 30 August 2010 (product ID 12345, socks) and from 1 September 2010 to 31 July 2011 (product ID 12345, washing machine). These things do happen!

How to sell to management: Include a graphic, for example, a dendrogram or other visualisation. Also, you can shorten the cluster descriptions and present them as a profile.

9.4 RECIPE 13: TO FIND AND DESCRIBE GROUPS OF CUSTOMERS WITH HOMOGENEOUS USAGE

In general, this recipe is a variation of Recipe 12. Recipe 12 is about products and Recipe 13 is about customers, so on one level you can just say that the objects are turned round. But handling single customers as a single line in datasets might give you more opportunities to add data at different levels of aggregation. So it seems to be a good idea to dedicate a recipe to this situation and to go into some detail.

Industry: The recipe is relevant to all sectors.

Areas of interest: The recipe is relevant to marketing, sales and strategic decisions.

Challenge: The challenge is that marketing as well as sales needs to have a more tangible picture or description of different customer or consumer groups. To provide these group descriptions, different solutions can be chosen:

1. Segmentation can be carried out by drawing on knowledge of the business and experience. For example, the segmentation may be based on gender, fixed age groups (like 18–24, 25–34, 35–44), etc.
    a. Advantage: Segmentation can be done easily, only limited statistical knowledge is necessary, and it is easy to explain to marketing managers. Also, it fits in with any pre-existing prejudices.
    b. Disadvantage: There is less chance to find unforeseen results and new knowledge; because the experiences are often not specific to a single company, it is very likely that you end up with the same segments as your competitor.
2. Segmentation can be carried out using a data-driven approach: the segmentation is based on statistical algorithms that try to find groups of customers with homogeneous behaviours.
    c. Advantage: There is good chance to find new target groups that were not obvious or easy to find. It is possible to find patterns that are specific only for this company, for example, target groups sharing the same age and gender but with very different behaviour. This group can now be separated from the others on the basis of this behaviour.
    d. Disadvantage: There is a need for statistical knowledge and statistical software. The ability to sell the results to marketing managers must also be very high.

Necessary data: The areas of data that could be used must contain information on all kinds of buying behaviour of single customers, advertisements and campaigns that could reach a customer and the general customer attributes.

Population: This is defined according to the actual problem and the business briefing. It has to be discussed whether you include all customers including active and inactive ones or not. In general, you need a splitting criterion for the segments that is not used to define the population used to learn on. So, for example, if we have used active and non-active customers to split up the population, then we cannot use 'general activity' as a criterion for the segmentation for either group because everyone has the same score for that variable.

Note that in cluster analysis, we have to reduce the population size; otherwise, the computational time becomes too large. We can reduce the population size either by using a splitting criterion such as gender or activity level or by random sampling. If we use random sampling, an important variable like gender or activity level is likely to dominate the clusters so that one cluster just contains the people in one part of the splitting criterion, and this is not a particularly useful outcome. Therefore, it is often more useful to use a splitting criterion to reduce the population size and produce two or more cluster analyses, one for each part of the splitting criterion.

Target variable: As we are dealing with an unsupervised learning problem, no target is needed.

Input data – must-haves: The must-have data includes:

- Raw data on historical reactions to previous marketing activities, for example, the customer recommended a friend to a publisher, or returned a question- naire, or used a voucher.
- Raw data on historical purchasing behaviour not just to previous cam- paigns, if any, but general historical behaviour. We include full details such as date, kind of product, amount, value, channel of purchasing, for example, online shop or local shop, salesperson and reductions such as buy one get one free.
- Raw data of the historical details of advertisements, promotions, visits by salesman and other communications sent to the customer.
- Cost per customer for every marketing and sales activity (marketing costs).
- Cost/margin for products.
- General customer attributes such as gender, age, occupation, salary and education.

Data mining methods:

- Frequency and mean values
- Clustering
- SOMs

How to do it:

Data preparation: In this example, the customers are the cases. In a similar way to other recipes, it is necessary to aggregate and transform the raw data. You should follow the rule of having one line of information storing several variables per individual customer. It is possible to use data marts that have the data in a suitable format.

Depending on the number of customers who are available, a sample is a good solution. For example, if your database contains 1.5 million customers, this number might be too large to get a clustering result back from the computer in a reasonable work time, so it is a good solution just to learn on a sample of maybe

30 000 customers. The number of variables should be limited according to business considerations, aggregation and the fact that most of the clustering algorithms can only handle a given number of variables to calculate the clusters.

Business issues: It is very likely that the results will be used as a first step for decision making and strategic planning, so the technical considerations regarding the implementation might be small in the beginning. But nevertheless, there might be a need to mark a single customer with the cluster number of the cluster they belong to, and so this should be done in readiness. Before you start the next steps, think about the potential implementation. Any step or manipulation that you know cannot be done when you implement the rules or cluster definitions should not be used in the learning phase.

Transformations: As always with modelling, the same kind of transformations and aggregation should be done to generate the variables that can be used in the cluster algorithm.

Analytics: We consider the three alternatives:

Experience-based approach of segmentation: To carry out this approach, first set up the rules for the experience-based segments, for example, one segment may be women aged 18–34 years old. Then group people in different segments with all their associated datasets and variables.

An example of potential segments is shown in Figure 9.6.

Segment	Age Group	Gender
S1	0–17 years	Female
S2	18–34 years	Female
S3	35–54 years	Female
S4	55–74 years	Female
S5	Older than 75	Female
S6	0–17 years	Male
S7	18–34 years	Male
S8	35–54 years	Male
S9	55–74 years	Male
S10	Older than 75	Male

FIGURE 9.6 Example of segments.

This approach uses frequency, ranks and means to calculate numbers that might help to describe the segment and its behaviour in detail. Some examples are:

- Number of people in the individual segment
- % of total customers
- Average revenue last year
- Average revenue in total
- Three most relevant products
- Three most common occupations
- Three most common residences
- % of people with Single Sign-On (SSO) Registration
- Most relevant social network
- ...

The question of the kind of numbers that are relevant depends on the individual business task and available information. The numbers add to the clarification as shown in Figure 9.7.

Data-driven approach of segmentation: This approach uses cluster algorithms or SOMs. Several algorithms are implemented in data mining tools.

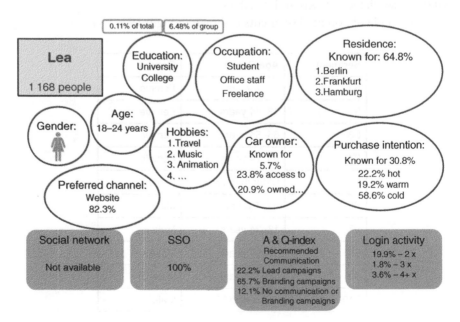

FIGURE 9.7   Potential ways to present a cluster (named Lea).

Before using them, it is important to check which size of dataset will operate or if there are any given limitations. Based on that restriction, it might be useful to reduce the original dataset by sampling and to reduce the variables to those which might be potentially used in the cluster description that will be presented to the management board. A hint of which kind of variables might be more relevant to separate different groups of customers can also be generated by a closer look at the list of important variables given in a predictive modelling project that you or one of your colleagues might have done before.

As a rule of thumb, the dataset should contain fewer than 40 000 rows and fewer than 250 variables to reduce the calculation time of the algorithm to a reasonable size. But with increasing computer power, this limit might expand.

In general, it is recommended to use the default settings of the data mining tool to start the clustering, especially if you do not have any prior knowledge on the amount of potentially existing clusters or other knowledge that might be useful to speed up the process.

Evaluation and validation: Based on the fact that cluster algorithms belong to the group of unsupervised learning, there is no additional process of validation than the check points delivered by the data mining tool. But nevertheless, it is still important to cross-check the results according to whether they make good business sense.

Implementation: Both kinds of segmentation can be implemented on really big amounts of data just by using the defined or detected rules:

For the experience-based approach, it is quite easy – just use the definitions that are used to build the segments.

For the data-driven approach, it is a little bit more complicated: If you can use an application function in your data mining tool that allows you to use a different dataset for the application than the one for training, it is quite easy. Just make sure that the new dataset includes the same variables and that the variable measurement is similar to that in the learning dataset; then it is possible to mark the dataset with cluster membership so that each customer is allocated to one of the detected segments.

In case your data mining tool does not provide such a function, then you have to work out the rules of implementation yourself. There are two different approaches: one is to generate the rules and the other is to develop the predictive models for each segment.

To generate rules, calculate the mean for each continuous variable in each segment and its accompanying confidence interval; for categorical variables, find the most common category in each segment. Then construct a set of rules to decide which segment to allocate each customer to. An example of these rules is as follows:

For each customer, if age is within the confidence interval for segment 1 and if the gender is in the modal category and if the home address is in the modal region, then allocate the customer to segment 1.

Clearly, this *ad hoc* approach can lead to some customers fitting into a number of different segments and other customers not fitting into any segments. The practical solution is to omit customers without a segment and allocate customers with multiple options to one segment at random. This is a pragmatic approach, and so if there are a lot of segment-less customers, then the process of allocation has to be repeated with wider confidence intervals.

If there are a lot of important variables defining the clusters, then the predictive model approach is recommended. To do the predictive modelling, work with the sample that was used for the cluster analysis and create a binary target variable which is 1 for segment 1 members and 0 for everyone else. Then create the model using the important variables from the cluster analysis. You can then apply the predictive model to the whole population and obtain a probability of having target = 1 for each customer. The process is repeated for each segment. We now have a set of probabilities of segment membership, one for each segment, and we can allocate the customer to the segment for which they have the highest probability of membership. Usually, this works well and may be faster than trying to generate rules that discriminate successfully.

Most of the time, an application on all (or new) data is not necessary, because the cluster results are just used strategically and for planning purposes. It helps to sell the results to management if meaningful names are worked out for each cluster, for example, young professionals or retired people.

9.5   RECIPE 14: TO PREDICT THE ORDER SIZE OF SINGLE PRODUCTS OR PRODUCT GROUPS

This problem can be solved as a predictive model. You can follow the recipes in Chapter 8 in general but with some amendments.

Target variable: Instead of the target being a binary variable (whether the customer responded to a particular offer or not), in this recipe, the target will be continuous or at least categorical. The target variable represents an aggregation of buying behaviour. You have to decide whether a detailed (continuous) target will add a real business benefit or not. If it is more important for the business to know whether a customer is likely to buy 'a few' items of a specific product than it is to

Customer ID	Model (0 item)	Model (1 item)	Model (2 item)	Model (3–5 item)	Model (6–10 item)	Model (>10 item)	Winner
3456	0.45	0.89	0.20	0.17	0.21	0.08	Model (1 item)
7890	0.36	0.65	0.83	0.42	0.35	0.20	Model (2 item)
1246	0.86	0.34	0.33	0.25	0.17	0.18	Model (0 item)
2378	0.23	0.43	0.45	0.43	0.65	0.67	Model (>10 item)

FIGURE 9.8 Example showing scores for different models.

predict the exact number, then the target should be classified into a categorical variable rather than being a continuous variable.

Data mining methods: If you decide to have the target as a continuous variable, then you can use multiple linear regression or another method suitable for a continuous target. If the target is classified, then logistic regression analysis can be used or you can develop separate models for each class as in Chapter 8.

Transformations: We strongly recommend that the input variables are transformed as described in Chapter 8. Using the binning or quantile methods helps to stabilise the models, especially if the environment where the input data is collected is constantly changing or if the data is collected directly from a legacy system. As a rule of thumb to decide whether the input data should be transformed to a classified variable or to keep it continuous, if it is a common pattern in your data that mean and median differ a lot, then classification of the input variables is recommended to keep the model more stable.

Implementation: If the target is classified, then you can develop a model for each class (just following Chapter 8). Each of these 'class' models should be deployed on the customers, and, for each customer, the model that gives the highest score wins (see Figure 9.8).

Notice that it is not always absolutely clear which model wins. For the last customer in the table, customer ID 2378, the only thing you can say is that it is likely that the person will buy in general and that it might be more than six items.

9.6 RECIPE 15: PRODUCT SET COMBINATION

This recipe is a common variation of Recipe 11. The target has changed from communication channels to products. The central issue is the definition of the product. It does not sound as if it is a big problem, but it can be quite difficult

FIGURE 9.9   Ten bottles of water, all the same?

depending upon your point of view. Consider the following definitions with reference to Figure 9.9 and the different viewpoints:

Logistic/warehouse: Each bottle has its own product ID and storage place in the warehouse. ⇒10 different products

Sales and marketing: Each bottle has its own product ID. It may belong to different brands, have a different size and a different price. ⇒10 different products

Customers: It is mineral water but customers may differentiate between natural and sparkling or between brands or between price levels. ⇒So the number of products depends on the context and how a customer might see the product; therefore, it could be from 1 product to 10 products.

From the point of view of analytics, you should decide on the aggregation level (product ID) to reflect the customer view depending on the business aspects, the available data and the goal that should be reached by the analytics.

It might be useful to do the analysis at least twice: First time to analyse on a highly aggregated level like 'mineral water', 'beef', 'vegetables', 'sweets', 'cereals' and 'soft drinks'. This helps to find out what the common combinations are and how big they are in total. Support and confidence are good indicators of the importance of the combinations as well. For the second time to analyse on a different level, for example,

Mineral water: Sparkling and natural

Versus

Beef: Neck, chuck, brisket, flank, rib, sirloin, tenderloin, top, round and shank

Versus

Vegetable: Mushrooms, carrots, spinach, peas, eggplant, tomatoes, potatoes and beans

As you may appreciate, the number of combinations increases dramatically and each potential combination might be small, or the support and confidence may be too small to indicate a trustable product setup.

So the most difficult issue in this recipe is to decide on the right level of aggregation of the products to generate stable results on the one hand and detailed level usable results on the other. A higher level of aggregation is better for generalisation because it is very likely that product details may change.

9.7 RECIPE 16: TO PREDICT THE FUTURE CUSTOMER LIFETIME VALUE OF A CUSTOMER

Industry: The recipe is relevant to all sectors.

Areas of interest: The recipe is relevant to strategic decisions, marketing, sales and control.

Challenge: Calculating the Customer Lifetime Value (CLV) at the level of existing customers or addresses known in the past is a common activity. However, the challenge here is to include a prediction of future customer behaviour. The prediction of future customer value makes all the difference because only this guides you to the right strategic decisions in customer development. The traditional way of using past data could mislead because it just deals with the past behaviour and does not include the potential forthcoming behaviour of the customer (see Figure 9.10).

So the challenge is to calculate the future value of a customer by combining the results of a prediction model that predicts the future affinity to order or not to churn (if you are in the subscription business) and traditional CLV models. So we have to do two steps: first work out a predictive model that enables us to estimate the future behaviour of a single customer, and secondly

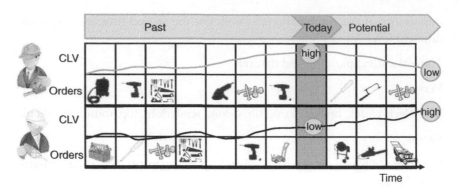

FIGURE 9.10 CLV development.

based on the prediction, we calculate the net present value for each single customer. As an example, consider the case where you have to predict the affinity to order; suppose you are a producer of machines for craftsmen and you sell directly via a catalogue, online shop and salesmen.

Necessary data: The areas of data that could be used must contain information on all kinds of advertisement and campaigns that could reach a customer including all associated costs and the general customer reaction not just the direct response to one advert or communication.

Population: The population is defined according to the actual problem and the business briefing. It has to be discussed whether you include all customers including active and inactive ones or not. If for older and inactive customers the data and especially the cost information are unavailable and cannot be estimated, it is better to concentrate on the active customers.

Target variable: The target is a binary variable such as 'buying' or 'not buying' in a specific time slot. For some companies especially in B2B, it makes sense to count successful recommendations as Target = 1 as well.

Input data – must-have: Must-have data includes:

- Raw data on historical reactions to previous marketing activities, for example, the customer recommended a friend to a publisher, or returned a questionnaire, or used a voucher.
- Raw data on historical purchasing behaviour not just to previous campaigns, if any, but general historical behaviour. We include full details such as date, kind of product, amount, value, channel of purchasing, for example, online shop, local shop, salesperson and reductions, for example, buy one get one free.

- Raw data of the historical details of advertisements, promotions, visits by salesman and other communications sent to the customer.
- Cost per customer for every marketing and sales activity (marketing costs).
- Cost/margin for products.
- General cost per customer for handling.
- Try to get the information as complete as possible; if some information is only available on a general level, try to break it down to customer level.

Data mining methods:

- A simple prediction of future buying behaviour (e.g. time series model).
- *CLV calculation*: In the simplest form, the net present value method can be used to determine the lifetime value. In this method, values are input for the expected customer proceeds, e_t (based on sales, cross-selling revenue, etc.), and the customer-specific costs, a_t (mailings, advice, investment, etc.), for each time period, t, in the expected duration, T, of the relationship. The costs are subtracted from the proceeds and discounted using a discount rate, i, and then summed over all the time periods:

$$CLV = \sum_{t=0}^{t=T} \frac{e_t - a_t}{(1+i)^t}$$

How to do it:

Data preparation: In this example, the customers are the cases and you just try to summarise both objects e_t and a_t for each time period *t*.

Business issues: Before you start the next steps, think about the potential implementation. Any step or manipulation that you know cannot be done when you implement should not be used in the learning phase.

It is very likely that the result will be used as a first step for decisions and strategic planning, so the technical considerations regarding the implementation might be small in the beginning. But note if the CLV-based strategies are implemented, there is an upcoming need to do the CLV calculations fully automatically and regularly.

Transformations: As in other predictive modelling questions, the same kind of transformations and aggregation should be done to generate the customised proceeds and costs for each past time period.

Analytics:

Pre-analytics: You should use this phase to control whether the chosen aggregation of your objects should be the customers or customer groupings or the way that purchases are accounted for. For example, should purchases made in the morning or in the afternoon count as being in the same day or should another system be used where a purchase is only counted when it is paid for? So at this stage, it is important to check that your numbers are all fitting in the same reporting universe, in other words that you have sound operational definitions.

Model building: The CLV for each customer is calculated using the formula provided earlier. For each customer, start with their individual setup costs from the time they began their business relationship with you. Then calculate costs and proceeds for all periods with real numbers out of the system. In many cases, a year is a suitable time period, t. Usually, data is used for each year that the customer was in the relationship with the company, which could be before they actually purchase if there is a long lead in, for example, companies selling cars, houses or new kitchens.

At the end, you will get the actual CLV without the forecasting aspect. The cost or profit of the customer is just added to the company's business results. Even if you stop here, these results already have great value for your company and will help to improve your CRM strategies.

For the prediction of the future customer behaviour, we recommend using a time series model based on revenue and advertisements over the past time periods. To get a result for the potential and expected future CLV of each single customer, you should use the estimated values in the formula for each future point in time. In most business areas, a forecast of three to five years ahead is fine.

Note that a 70-year-old will have a greater CLV than a 35-year-old but may not be such a good future customer. Factors such as biological age, however, may not have such a great impact if the forecast period is short. The impact will also vary with product, for example, age is unlikely to impact on buying food but may impact on investment purchases. Changes in costs and proceeds over time will be part of the variation in the data used in the time series modelling and so should carry through into the forecasts. However, it is also possible to incorporate explanatory variables explicitly in time series modelling. Note that a good customer may not necessarily remain good (see Figure 9.11).

Evaluation and validation: We need to check both the prediction model and the calculated CLV. Apart from making sure that your calculation is correct on a technical level, the only way to validate the models is to cross-check the results with figures from other reports (your own and those of any other companies that you can find) to see whether the types of people that you predict will be valuable customers usually turn out so to be.

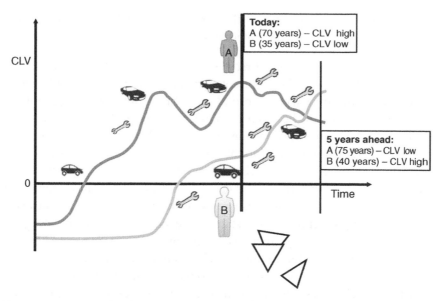

FIGURE 9.11 CLV development. Note that good today may not be good tomorrow.

Do the rules of the time series models fit with the business rules? If there are rules and patterns in the model that we cannot understand, and neither you nor anyone else knows why they are there, this suggests that the model is over-fitted and something is wrong with it. Apart from cross-checking with other numbers, no additional validation of the whole CLV is necessary.

Implementation: There are two particularly important points to note:

The results are mostly used for strategic decision so there are no technical barriers to be aware of.

In case the algorithm is implemented, make sure that it enables you to replace the forecast by real numbers in the model and do a new forecast after the period is over.

Hints and tips: You will get the CLV for customers as a continuous value. Additionally, you should group the CLV into classes with the help of statistical or business rules. For example, the CLV for a particular customer may be 200 000 €, and overall the customers may have CLVs ranging from 10 000 to 2 million €. We would look at the frequency distribution to divide the customers up or use business rules developed over the history of the company from experience or from a wish list. For example, any customers with CLV over 1 million € may be counted as one category, or historically, they may be further grouped into a number of categories. Finally, we then end up with a table showing the % of customers with CLV in different categories.

How to sell to management: As a first step, it might be enough to include simple groupings of the CLV in the reports, for example, customers with a 'big loss', 'small loss', 'hardly any balance (around zero)', 'small profit' and 'big profit'.

A good overview can be given by presenting the results in the form of a matrix with CLV categories on one axis and predictions on the other axis. In the so-called Boston matrix format, there are four groups in a 2 × 2 contingency table with high and low CLV in rows and high and low prediction in columns. Enter the numbers or % of customers in each part. The customers in the four parts of the table are: cash cows but no further development likely (high CLV, low prediction), rising stars with good value at present and good potential (high CLV, high prediction), uncertainties for whom we are not sure what will happen in the future (low CLV, high prediction) and poor dogs with no hope now and no hope of any improvement in the future (low CLV, low prediction). Such tables were originally used for market share and growth but are now used more widely.

We are calculating a CLV for each person so we can group the values by region and salesman or customer types or by association, for example, we can classify by use of hotel group or holiday type. This would indicate which other companies it is good to team up with.

You should note that the calculation of the CLV will start a major discussion in your company. In our example, it is very likely that it might have an influence on the management of the salesmen.

Learning from a Small Testing Sample and Prediction

To illustrate these recipes, we will use data from the JMP challenge presented at the ENBIS 2012 conference in Ljubljana. The methods recommended are regression and decision tree analysis, and other possibilities include neural networks.

10.1 RECIPE 17: TO PREDICT DEMOGRAPHIC SIGNS (LIKE SEX, AGE, EDUCATION AND INCOME)

Industry: The recipe is relevant to everybody who needs a clear customer demographic profile to improve their business, for example, companies publishing banners and other groups preparing specific advertising offers.

A Practical Guide to Data Mining for Business and Industry, First Edition.
Andrea Ahlemeyer-Stubbe and Shirley Coleman.
© 2014 John Wiley & Sons, Ltd. Published 2014 by John Wiley & Sons, Ltd.
Companion website: www.wiley.com/go/data_mining

Areas of interest: The recipe is relevant to marketing, sales, online promotions and strategic decisions.

Challenge: The challenge is to get a clear picture of the demographic distribution of customers and prospects or to replace missing values in the database (referred to as imputation). The latter is the more common situation. Even if fields like sex, age or income are available in the datasets, it quite often happens that large parts of the data are missing because the process supplying the values did not ensure that these fields were completed. But this sort of information is needed to get a clear picture, to make the right strategic decisions or to sell the information to others to optimise their communication. To build the prediction models, it is necessary to know at least part of the real data (or data that can be defined as real) for the person. This data can be extracted from the database (from those with fully filled datasets) or can be generated through questionnaires filled out by some of the relevant people. It is important to make sure that the people with full datasets or those who get the questionnaires somehow represent the whole population (see Figure 10.1).

Necessary data: The areas of data that could be used must have some direct relationship to the customer behaviour or must come directly from the customer (e.g. data directly from the purchasing process or marketing activities or out of log files that have been used).

Population: The population is defined according to the situation, whether it is an imputation problem and quite a large part of the data includes all information so you can use it for learning or whether you are faced with the situation that data with full information (full datasets) are generated by questionnaires given to customers or users. These two problem areas are comparable if the

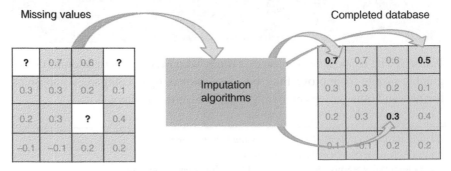

FIGURE 10.1 Obtaining a full database.

questionnaire can be linked with other data through a key variable or other variables such as name and address. Those people with full information (full datasets) are used for learning.

We distinguish our sampling frame population which consists of the people with full or partial information from the population of people upon whom we apply the knowledge gained from the modelling.

Target variable: The target variable can be age, sex, income, etc. The easiest and most stable way to deal with such variables is to convert them to groups of binary variables by using bins. For age and income, you might think let's use the real measurement and do the prediction on continuous variables. The answer is quite simple; you can do that if you have a very solid information base, and age and income are needed as detailed information. But actually using the detailed information might give results that are less stable than those obtained by grouping the variables into meaningful bins. These bins should include your business knowledge and should be compatible with the groups that are typically used by the marketing department. The bins may vary in size and may depend on the cultural background. If the data to learn on is generated by a questionnaire, it is very likely that the bins are already given by the questionnaire.

As an example, age can be grouped as follows:
1–13 years
14–18 years
19–25 years
26–35 years
36–45 years
46–55 years
56–65 years
66–75 years
Older than 76 years

Monthly income can have following bins:
1–500 €
501–1000 €
1001–1500 €
1501–2000 €
2001–2500 €
2501–3500 €
3501–5000 €
More than 5000 €

If you use these bins, you will end up with nine different targets for age and eight different targets for income, each of which can be modelled. As a rule of thumb, with fewer bins, the models might be better but you lose information.

Input data – must-have: If your imputation has to be done on datasets where you have access to offline behavioural data, then the following data will help:

- Raw data on historical reactions to previous marketing activities, for example, the customer recommended a friend to a publisher, or returned a questionnaire, or used a voucher.
- Raw data on historical purchasing behaviour, if any, or general historical behaviour. We include full details such as date; kind of product; amount; value; salesperson; reductions, for example, use of buy one, get one free; and channel of purchasing, such as department store, online shop and local shop.
- Demographic and more data such as customer postcode, longevity of customer, preferred channel, channel of first contact, source and family size. If we are considering B2B, then also consider the company size; company section, for example, steel, oil or coal; and branch location.
- Historical changes at the customer level, for example, changing from one customer worth category to another, moving addresses or changing marital status.

In case your imputation problem is based on online data, you must use the log file or data extracted out of the log file as input data. Also incorporate any details of online facilities used by the customer in login areas in your website, for example, looking for new, upcoming cars, books or other products or services assuming they are featured on your website.

Data – nice to have: This data includes:

- Raw data of the historical details of advertisements, promotions and other communications sent to the customer
- Raw data on all complaints that are not related to the buying process itself, for example, asking for status of commitment in terms such as 'when can I stop my contract?'
- Whether customers recommend other people
- Raw data on historical paying behaviour, for example, credit card, late payments, reminder letters, paying by cash and instalments or full amount

Data mining methods:

- Decision trees
- Regression, especially logistic regression
- Neural networks

How to do it:

Business issues: We need to identify any known exceptional periods or variables, for example, regular marketing activities that promise customer reductions or discounts based on age or sex. If possible, Target = 1 customers created at that time should be excluded from the learning system because it is quite likely that they do not give the full picture as they were only attracted by the temporary benefit. These customers can sometimes be dealt with by adding an indicator variable or if the period is small, excluding them.

In most countries, there are clear correlations between first name and sex or first name and age. Sometimes, there are also correlations between social background and first name. If you assume that these kinds of correlation exist in your data, you can use them for prediction as well. Sometimes, this kind of prediction is good enough to solve the problem.

Figure 10.2 shows an easy way to estimate age and/or sex on the basis of the first name.

In some countries like Germany, most first names and/or the first and middle name combinations give a clear hint on the sex and often on the age range of a person as well. As background: there are fashions which influence the likelihood of a name being given to new born child; it is not random; it is mainly influenced by trends or 'crazes', family traditions, movies and parent education, in that order. Following this route, you can calculate a distribution of ages for each first name.

To do this, you must analyse those of your addresses with a full combination of first name and age. If you plot the histogram of ages for each first name, you

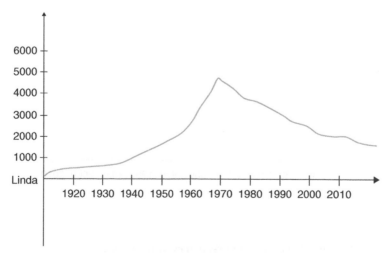

FIGURE 10.2   Allocation of first names with 1 peak.

will find a single peak for most first names, sometimes with quite a small variance (this happens for names that were very fashionable when they were given to the baby). Some of the first names will produce a double peak as people with this first name are usually quite young or quite old. Some names have no obvious peak age (see Figure 10.3).

In the case of the single peak distribution, the age can be estimated through the mode or the median. In the case of the distribution with no obvious peak, you can use the mean or the median (see Figure 10.4).

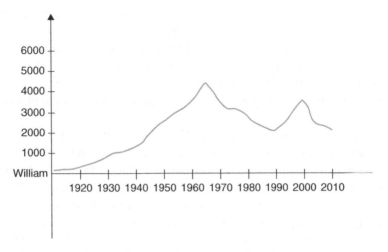

FIGURE 10.3 Spread/allocation of first names with 2 peaks.

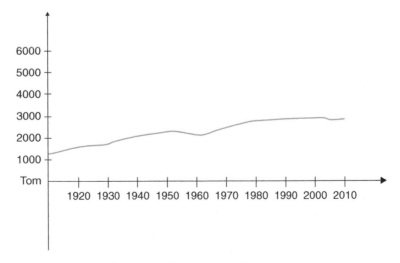

FIGURE 10.4 Spread/allocation of first names with 0 peaks.

First Name	Count	Median Age/Current Year	Year of birth
Berryl	8	64	1949
Brian	5	45.5	1968
Burt	19	42	1971
Britney	8	30	1983
Brittany	24	24	1989
Chloe	11	24	1989
Edgar	17	70	1943
Elmer	7	73	1940
Jake	5	23	1990
Kaitlin	13	24	1989
Kaley	5	24	1989
Katelyn	12	22.5	1991
Kayla	10	23	1990
Mallory	11	24	1989
Max	8	72	1941
Wendell	8	70	1943
Whitney	10	24	1989
William	19	43	1970

FIGURE 10.5   Look-up table for first name and estimated age.

If you have a distribution with two clear peaks, it is necessary to think of another additional indication that helps to decide which estimate of age is the most suitable for the person.

So you can create a look-up table with first name and estimated age. To be on the safe side especially if you plan to use the look-up table for quite a long time, it is better to store the established year of birth rather than the actual age (see Figure 10.5).

Similarly, you can use the full first name–sex combination to predict whether the first name is more likely to be for a female or for a male.

In case your imputation problem cannot be solved that easily, it can be seen as an ordinary prediction problem using the known full data with the variable that needs imputing as the target and the existing behavioural data as input. Following that route, you have to check whether transformations and other steps to prepare the data are necessary and may improve the results.

Transformations: The purpose of transformation is to stabilise the final models. For this, we need to replace missing values, exclude outliers (or ameliorate them), smooth the data, Normalise to use methods requiring Normality and reduce the effects of changes in mean and variance. Note that in the decision tree method, the transformation can be done by the algorithm (usually grouping into a small number, say, 10, of groups). Transformation is very important to improve results; otherwise, models can be over-affected by the vagaries of the data mentioned earlier.

The next part depends upon whether you already have a marketing database with information for a single customer all in one long row or not. If not, then you need to convert the data 'from small and long to short and wide'.

Marketing database: You now need to think of suitable transformations for the input data. We recommend using the quantiles or bucket method if you decide to have an ordinal target or standardisation if your target is continuous.

Quantile method: Recall that for this method, we first order the column of data according to the data values and then divide the rows into six quantiles. We note the quantile borders; then, using an 'if then else construction', we transform the real data into six classes between 0 and 5. As an important aside, note that typically for some input data we can have more than 95% zeros. In this case, we could first check if more than 5/6 of the data values are zeros, and if they are, then we could allocate the 6th quantile to 1. However, if we want a consistent method for all the many variables, we stick with allocating the 6th quantile as 5, although we can, of course, use a different method for each variable.

Bucket method: Transformation can also be carried out separately for each variable 'by hand'. It is probably better to use an algorithm though as the benefit gained from a more tailored approach is not usually worth the effort.

Note that the chosen transformation must coincide with the unwritten rules of the business, for example, that customers from rural areas are grouped together, because some variables like ordered postcodes do not give a good grouping. In the case that single variables are filled with numbers representing codes, you have to do your grouping for this variable by hand or create dummy variables out of it.

Standardisation: This procedure is helpful for continuous variables as it avoids differences in scale, making the models unstable.

Another typical problem is missing values like those shown in Figure 10.6.

Analytics:

Partitioning the data: Patterns found in a representative sample of the data are usually a good indicator of patterns in the whole dataset. We take a stratified random sample based on the target variable to ensure that we have enough people in each category of the target variable. If the target is a binary variable, then the sample could be made up of approximately 50% with each value of the target. This helps to keep the resulting model stable, whereas if the sample is mostly Target = 0, then the model will tend to fit the Target = 0 people better

	target_female	target_male	target_A geto29	target_A e3039	target_Ag e4049	target_Ag e5059	target_Age60	target_job	target_salary_2500	target_hou shold1	target_hou shold2	target_hous hold345
55448 fb_c	0	1	0	0	0	1	0	•	•	1	0	0
55449 bd_c	0	1	0	1	0	0	0	•	•	0	1	0
55450 l_c	0	1	1	0	0	0	0	1	0	1	0	0
55451 la_b	1	0	1	0	0	0	0	0	•	0	0	1
55452 bd_b	1	0	1	0	0	0	0	0	•	1	0	0
55453 bc_b	0	1	0	0	0	0	0	1	0	1	0	0
55454 b	1	1	1	0	0	1	0	0	1	0	0	1
55455 l_a	•	•	•	•	•	•	•	•	•	•	•	•
55456 pf_c	1	0	0	0	0	0	0	•	•	0	0	0
55457 l_b	0	1	1	0	1	0	0	1	0	1	0	0
55458 _c	0	1	0	0	0	1	0	•	•	0	0	1
55459 46_b	1	0	1	0	0	0	0	0	1	0	0	1
55460 4_a	0	1	1	0	0	0	0	0	0	0	0	1
55461 7_c	1	0	1	0	0	0	0	•	•	0	0	1
55462 j20_c	1	0	1	0	0	0	0	1	0	0	0	1
55463 2_b	0	1	1	0	0	0	0	1	1	0	0	1
55464 6_a	0	1	1	0	0	0	0	1	0	0	1	0
55465 _a	0	1	0	0	0	0	0	1	0	0	0	1
55466 6_a	•	•	•	•	•	•	•	•	•	•	•	•
55467 _c	1	0	0	0	0	1	0	0	1	0	0	0
55468 c_c	1	0	0	0	0	1	0	0	1	0	0	1
55469 _b	1	0	0	0	1	0	0	1	•	0	0	1
55470 j6_c	0	1	1	0	0	0	1	0	1	0	0	1
55471 j3_a	0	1	1	0	0	0	0	0	•	0	0	1
55472 pd_a	1	0	0	0	0	0	1	0	0	0	1	0
55473 _a	1	0	0	0	0	0	1	0	0	0	1	0
55474 j76_c	1	0	0	0	0	0	1	1	1	1	0	0
55475 _a	0	1	0	0	0	1	1	0	•	0	1	0
55476 _a	1	0	0	0	0	0	0	0	•	0	0	1
55477 _c	0	1	0	0	0	1	0	1	0	0	0	1

FIGURE 10.6 Missing values to be imputed by predictive modelling.

than the Target = 1 (usually the active customers) people. Also, we want to understand the specific attributes of those with Target = 1, and so we need to have a good number of them in the sample.

People with Target = 1 will tend to have fuller datasets including all the information about their purchasing behaviour. If the population has a smaller disproportionate number of Target = 1 people, we may have fewer full datasets. We will determine the sample size from the number of customers with full datasets. Typically, we take three samples at random from the Target = 1 people and also take three samples of the same size from the Target = 0 people. Resulting sample sizes tend to range from 600 to 60 000. As a rule of thumb, if the customers are very active, you can work with a smaller sample, but if they are less active (i.e. buying only once or twice a year, or relative to what is expected in the business they do not buy very often), then you need a larger sample. This is similar to the usual rules for sampling that you need a bigger sample when you have less information.

Pre-analytics: This part of the analytics enables us to screen out some variables, for example, variables that have zero value or are all one value. Feature selection can be done with both the original variables and the transformed variables or with just the transformed variables depending on the number of variables and the degree of transformation that was required. It is best to keep original variables and transformed variables available in the dataset.

Feature selection using chi-square testing requires that the variables representing the features have a small number of categories. This indicates that it is best to use variables transformed by the quantiles method. Feature selection of variables transformed by the mid-range standardisation, or the original variables with many different values, uses methods including parametric or non-parametric correlations, principal components analysis, factor analysis and partial least squares.

We continue with analysing the selected, significant variables. The input variables are likely to be highly correlated. However, all variables are potentially important, so we cannot ignore variables just because they are highly correlated. Using the chi-square tests and other appropriate methods helps to retain the important variables for further analysis.

Model building: The model building preference is logistic regression for binary targets and linear regression for continuous targets. Normality is important for logistic regression, and the previous transformations help to make the input variables suitable.

Evaluation and validation: This is carried out by using the model on the test and validation samples and comparing observed and predicted outcomes in a confusion matrix. For better control, we can use a detailed report or gain chart that is free from cut-offs so that you can see how the scores for each group of customers developed. ROC and lift charts can also be used.

You need to check whether the rules found fit with the business, for example, the variables in the model should coincide with 'gut' feeling and you should feel at ease with the variables that have not been included. If there are variables in the model that we cannot understand, and neither you nor anyone else knows why they are there, this suggests that the model is over-fitted and something is wrong with it.

We also have to look for special effects which show up via the model which we were not aware of or had been underestimated or forgotten. These could be specific to particular features, for example, which only occurred in one year, like big promotions for anniversaries or for the 150th year of the company, or some other abnormal behaviour like everybody born in the same year as the company or brand gets a major discount. We then note these aspects of the model and allow for them in the predictions. Also, very importantly, consider if the model is useful for the business.

Implementation: Here, we address the original statement of the recipe, that is, how to predict demographic signs.

We need to consider how to implement the model or the rules or look-up tables on the actual data: are there any special needs or exceptional variables that cannot be transformed, or are there any programming issues?

Note that the distribution of the customers in the sample should be similar to that in the actual population, so that the learning is representative of the actual data.

Hints and tips: If you have lots of variables, you can replace the original variables by the transformed variables. In difficult cases where you are unsure of the model quality or it does not fit with the unwritten rules of the business domain or you get the impression that the model fits the sample but when it is generalised it will not fit so well, then try model building using several samples and compare the results, taking the variables forward that occur most often. This is similar to the method known as random forest in which a forest of decision trees is generated using different samples and the models are automatically compared to show which variables occur most commonly in the decision trees.

You can also gradually build on your experience of dealing with similar situations or customers and let this influence your final model choice. Note that the important issue is whether the model is validated and if it is useful when implemented.

How to sell to management: Normally, management does not worry about the cost of data; however, if the cost of the data is an issue for management, the best way to deal with this is to point out that your analysis saves the money that would have to be spent if you tried to buy the information on the open market.

10.2 RECIPE 18: TO PREDICT THE POTENTIAL CUSTOMERS OF A BRAND NEW PRODUCT OR SERVICE IN YOUR DATABASES

Industry: The recipe is relevant to all companies who would like to know at an early stage of product or brand development whether the expected target population is big enough and whether the product is likely to be a success and where to find those customers who are likely to choose it.

Areas of interest: The recipe is relevant to marketing, sales, online promotions and strategic decisions.

Challenge: The challenge is to get a better estimation of the potential target groups and the upcoming sales numbers. It is sometimes not enough just to follow the classical market research approach especially if you plan to introduce the new product, service or brand to your existing customers or if you plan to use publicly available addresses for marketing campaigns. The general idea is to learn a prediction model based on a sample of people with access to the prototype or who have been interviewed about their views of the innovation and whether they think that they might be interested in it if it is publicly available in the future (see Figure 10.7).

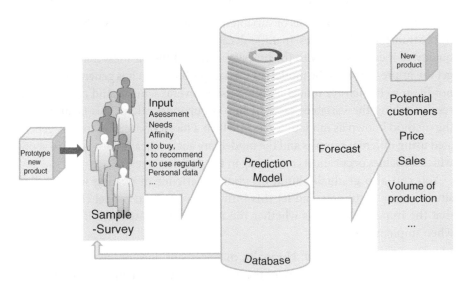

FIGURE 10.7 Typical application of prediction.

Necessary data: To be able to do the projection, you need data of at least three kinds:

1. Data from actual interviews/questionnaires or product reviews (including the individual judgments).
2. A dataset with additional information on the respondents; this information must be comparable to the dataset available during roll-out. If this is not given, there must, at least, be a mapping between that information and the information available in the dataset on which you will do the roll-out.
3. A dataset for roll-out or at least a good and valid data description if you have no physical access to the data file during the analytics stage.

Population: Depending on how the interviews and the additional data are collected, you may have enough full information to use these people directly to develop the model.

Target variable: The target variable should be generated from the conclusions of the interview or product review. It should be constructed from the categorised answers to questions like:

- How likely is it that you will buy the product when it is available?
- Will you recommend this product to friends?
- Do you think you will use this product/service regularly if it is available?
- Would you use this service if we charge for it?
- Give a fee/price that reflects the value you put on the service

It is important to try to reflect the consumer view when constructing the target. If the answer is given on a scale of 1–5 (1 = strong recommendation and 5 = no recommendation at all), you have to decide how to generate Target = 1 and Target = 0. For example:

1 = strong recommendation $\Rightarrow$ Target = 1

5 = no recommendation $\Rightarrow$ Target = 0

Ignore everyone else

If the number of Target = 1 and Target = 0 cases is big enough (at least more than 300 cases each), you do not need to use the other answers, and you have the chance to find a clear pattern during the modelling phase. If the number of cases on one or both sides is too small, you might add those who answered 2 or 4. But you should not use the answer 3 in this example because those people might not have an opinion on the subject at all. The only opportunity to use answer 3 as well is when you know by cross-checking of different given answers to other questions of your survey that answer 3 is just a hidden and polite way of saying 'no recommendation'.

So building a target might be quite difficult, and you have to cross-check the potential real opinion of the consumers. You must consider also the culture and social background of the consumers who have been questioned in comparison to the consumers on whom you will do the roll-out. Cultural differences can affect the respondent's willingness to give extreme views, and so a model learnt with one cultural background will not necessarily work on a different one. As you know in data mining, training data and roll-out data should be as similar as possible.

Note that all information used to create the target should be deleted out of the list of potential input variables.

Input data – must-have: If your model will be rolled out using existing data from your marketing database, then there are two different approaches:

The first approach applies if there is no common key or ID that links your interview results into your marketing database. In this case, you can only use those variables for the modelling that can be mapped somehow to your marketing database. For example, gender and age from your interview will map directly to gender and age in your marketing database. But other questions may be more difficult to match. For example, questions like 'How long have you been a customer?' may map into variables like 'customer since'. Questions after using the product like 'How many times are you likely to buy this product?' may be mapped to aggregated purchasing information on the product level. The process needs a lot of creativity, and it is important to map according to the business understanding; you may not be guided only by similar names, but there must also be comparable meanings. Any mapping between transformed and aggregated information on both sides needs to follow these rules:

- It must support the modelling.
- The transformed and aggregated variables must 'look' similar on both sides of the mapping.

The second approach is when a common key or ID exists. In this case, you have to check whether the answers to the non-target questions fit in with the recorded behaviour in the database; if they do, then all available data out of your database can be used for modelling. If the answers do not fit, then the first approach has to be used.

The data from the database that can be used includes the following:

- Raw data on historical reactions to previous marketing activities, for example, the customer recommended a friend to a publisher, or returned a questionnaire, or used a voucher.
- Raw data on historical purchasing behaviour, if any, or general historical behaviour. We include full details, for example, date; kind of product; amount; value; channel of purchasing, for example, department store, online shop and local shop; salesperson; and reductions, for example, use of buy one, get one free.
- Demographic and other data, for example, customer postcode, longevity of customer, preferred channel, channel of first contact, source and family size. If we are considering B2B, then also consider company size; company sector, for example, steel, oil or coal; and branch location.
- Historical changes on the customer level, for example, changing from one customer worth category to another, moving addresses and changing marital status.

Nice to have: The following data is nice to have:

- Raw data of the historical details of advertisements, promotions and other communications sent to the customer
- Raw data on all complaints that are not related to the buying process itself, for example, asking for status of commitment in terms such as 'when can I stop my contract?'
- Whether customers recommend other people
- Raw data on historical paying behaviour, for example, credit card, late payments, reminder letters, paying by cash, instalments or full amount

Data mining methods

- Decision trees
- Regression, especially logistic regression
- Neural networks

Business issues: There are no further business issues apart from those described in the input data section.

Transformations: If you plan to do the modelling on the questionnaire data including the necessary mapping as mentioned earlier, then the necessary transformations are indicated by the mapping process.

Analytics:

Partitioning the data: Depending on the number of cases in the dataset with Target = 0 and Target = 1, you can choose whether to partition your dataset into training and test datasets using random samples. Stratification to get a 50/50 relationship is not necessary unless the numbers with Target = 0 and Target = 1 are too unbalanced, for example, if Target = 0 is more than 95%. If you do not have sufficient information as regards the number of Target = 0 and Target = 1 people, then the rule of thumb is that if there are fewer than 2000 cases, it is better to employ cross-validation instead of using a partition into training and test for the purposes of validation. In most data mining tools, you can choose that as an option. If you have no access to cross-validation in your toolkit, then you should stay with training and test.

Pre-analytics: Depending on the specific situation, it may be necessary to carry out further transformations and aggregations to ensure the input variables are in a suitable form.

Model building: The model building preference is definitely for decision tree analysis; the method is very robust and will deliver rules that can be translated and used under different data setups, so the result can be transferred to other similar situations.

Evaluation and validation: Evaluation is addressed by considering whether the rules found fit in with the business and the results of previous market research activities in this area of interest. Also, very importantly, consider if the model is useful for the business.

Validation is carried out by using the model on the other samples and comparing observed with predicted outcomes in a confusion matrix. For better control, we can use a detailed report or gain chart that is free from cut-offs so that you can see how the scores for each group of customers developed. ROC and lift charts can also be used. Alternatively cross-validation can be used.

Implementation: In case the model was developed on the questionnaire dataset, we have to choose the way of mapping to apply the detected rules on our database. In case the model was developed on variables out of the database and just the target was chosen out of the questionnaire, the deployment of the model should be quite straightforward and comparable to other prediction problems.

Hints and tips: In addition to the general hints for predictive models, it is very important to document the mapping and all transformations and assumptions based on the questionnaire data.

How to sell to management: Highlight the predictive power and the opportunities to get selected addresses of potential hot leads for the new product instead of a general target group description as you get normally out of the market research.

10.3   RECIPE 19: TO UNDERSTAND OPERATIONAL FEATURES AND GENERAL BUSINESS FORECASTING

Data mining techniques can successfully be applied to smaller or simpler sets of data than those considered in the earlier recipes. Small to medium enterprises (SMEs) collect data for audit and tax purposes, and these can be used to understand and improve the running of the business. It is remarkable what insights can be drawn from such data, and this is demonstrated here for a small sample of data from an SME in the clothes retail sector.

Industry: This recipe applies to any business in the retail sector including SMEs where sales are made direct to customers from a retail outlet.

Areas of interest: The recipe is relevant for helping with strategic and operational decisions.

Challenge: The challenge is to identify ways to improve the business, bearing in mind that the amount of trade data is limited and that conditions change very quickly with time. Specific questions concern the numbers of customers, the numbers of customers per hour, the sales income, the sales income per hour and the sales income per customer as follows:

- What is the overall level of sales income?
- How does the level vary by day of the week?
- Has the level changed over the year?
- Does the level show seasonal effects?
- Can the balance of staff and income be improved?
- Can we predict well enough to avoid waste?

Typical application: These include identifying training opportunities by comparing sales achieved by different staff, confirming opening hours by calculating sales per hour on different days and planning for seasonal events by observing the increases in customers and sales around Easter and Christmas and changes between winter, spring, summer and autumn seasons.

Necessary data: Regular detailed data are required. The type of data collected varies by business. The more data that is available, the more analysis can be done.

Population: The population is the whole set of customers who have visited the retail outlet.

Target variables: As expected, the aim is to increase sales, and so this is the most important target variable. If data are not available on a customer basis, then the analysis can be based upon daily customer numbers and sales income.

Input data – must-have: For a basic analysis of sales patterns, necessary data includes dates, opening hours, daily sales, numbers of customers and staff rotas.

Input data – nice to haves: Nice to have data includes weather observations.

Possible data mining methods: As the data are very simple, basic data manipulation and graphical methods are used. These include calculating %, cross tabulations, bar charts and time series graphs.

Data preparation: The data are plotted and checked for outliers.

Business issues: Allowance has to be made for special days, such as public holidays.

Transformation: The data do not require any transformations.

Marketing database: The data are gathered into an Excel spreadsheet.

Partitioning the data: No partitioning is carried out because only limited data is available. It is quite common that trading conditions change a lot, and only the more recent data are consistent with future predictions.

Pre-analytics: A wide range of expressions may be used to describe the weather, and these can be grouped into a number of categories.

Model building: The data are summarised and illustrated with descriptive statistics; decision tree and regression models can also be useful.

Evaluation: The relevance of the analysis is considered to ensure it addresses the initial questions.

Validation: This is by face validation comparing the results with the common viewpoint.

Implementation: The analysis does not necessarily need to be repeated regularly, but the results are used to help with strategic business decisions. In this example, the results of the analysis confirm that the business is progressing satisfactorily, and no major changes in practice are needed at the moment.

Hints and tips: If data are limited, then it is important not to jump to conclusions. For example, if some members of staff only work on certain days and sales are low on those days, then the effects on sales income of staff and days cannot be separated. Special events and circumstances should be noted as these will affect the data.

How to sell to management: The data analysis must be presented in a report with clear messages outlined.

This recipe presents a strategic project typically carried out between management and data miner. It differs from the very large projects that involve operational specialists on one side and analysts on the other.

For the strategic project, we recommend carrying out the following stages:

Stage 1: Meet project owner and agree objectives; get data and do first interview with experts to collect domain knowledge which can be helpful for modelling.

Stage 2: Set up data, clean and check quality; transform, normalise, standardise, do basic statistics and order variables.

Stage 3: Report preliminary results and start the modelling process:
1. Discuss preliminary results with domain experts; check that all important variables are as expected or add more detail to them.
2. Start intensive modelling phase: create model, evaluate, validate, create a new model, evaluate, validate and repeat cycle until you find a suitable model.

Stage 4: Discuss final results and prepare any further implementation.

Stage 5: Confirm results of project and check on customer satisfaction.

These stages correspond to the Plan, Do, Check, Act cycle used in quality management and the Define, Measure, Analyse, Improve and Control (DMAIC) stages in Six Sigma process improvement.

The time period for all the stages depends on the application, the data and its quality but is usually a fairly quick turnaround to tie in with the business case, so of the order of weeks rather than months.

11

Miscellaneous

11.1 RECIPE 20: TO FIND CUSTOMERS WHO WILL POTENTIALLY CHURN

Industry: The recipe is relevant to publishers, finance, insurance, software, online services such as premium accounts of social networks and all industries with long-term contracts with clients.

Areas of interest: The recipe is relevant to marketing, sales and management.

Challenge: Some businesses are based on long-term contracts with less interaction between the parties, for example, contracts for most kinds of insurance

A Practical Guide to Data Mining for Business and Industry, First Edition.
Andrea Ahlemeyer-Stubbe and Shirley Coleman.
© 2014 John Wiley & Sons, Ltd. Published 2014 by John Wiley & Sons, Ltd.
Companion website: www.wiley.com/go/data_mining

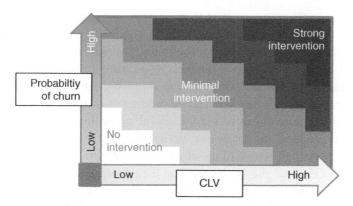

FIGURE 11.1   Intervention possibilities depending on churn and CLV.

and telecommunication, licensed software and daily delivered newspapers. In all of these businesses, it is pertinent to keep the customer or client in the system for as long as possible. Every customer that quits a contract too soon is lost money and is difficult to replace by a new one. It is an advantage for the companies to know in advance which customers are likely to quit and to work out ways to avoid it or to lower the probability. When customers quit or change supplier, it is referred to as customer churning (see Figure 11.1).

Necessary data: As in other recipes, order and contract data for the single customer are very important, but because of the type of business, this information may not reflect the customer behaviour as well as it does in the retail sector, and hence, it will not be sufficient to solve the problem of predicting which customers will churn. It is vital to get access to any kind of data from customer service or related online services and also if available to have data on the individual usage of the product. For example, how often does a customer use the software or the specific service, and is their usage increasing or decreasing? Customer payment habits prove to be very valuable and also the history of any advertising actions (online or offline) directed to the customers and their reactions to them.

Population: The recipe is relevant to all customers with at least one active contract at a specific day in the past. Make sure that there is enough time to measure churn between now and the day you have chosen. Churn rates reported elsewhere may help you to decide how long the time period needs to be to model the churn. If there are any seasonal aspects that are relevant to your business, these will also influence your choice. Everyone relevant for the analysis has to have at least one active long-term contract in place on the specific day and you need to clearly define the period during which you can usefully measure the churn.

Target variable: The target variable is whether the customer churns or not. However, you have to make sure that you have included the right behaviour as churn. We have to separate whether the churn is because the customer wants to stop the contract or the overall business relationship or whether the churn is because the company itself wants to stop supplying the original product and swop the customer to a newly launched product or contract. Churn may also come about because client was a successful up-selling action and the old contract was stopped and a new one was opened. You might wonder why all of these alternatives can be mixed up with customer churn. The reason is that the business process and the corresponding data footprints are similar in each situation; there will be a data case with an opening and an end date. Ideally, you will find additional indicators or variables in the dataset that will help to qualify the real targets. As a rule of thumb, you should project forward from the customer's viewpoint; does he or she stop the specific service completely, or do they continue with a similar or upgraded product? Another indicator might be that the closing date of the finished product or service is actually the same date as the start of a new contract. The target needed to develop a churn model is whether or not the customer quits without immediately starting a new contract.

Input data – must-haves: The following data are must-haves:

- Order and contract data (existing and former contracts or bought products)
- Data on received invoices
- Customer services/online services (like 'my account')
- Data on advertising activities (online and offline including newsletters)

Input data – nice to have: The following data are nice to have:

- Payment data
- Usage history of services, software or knowledge

Possible data mining methods: Comparable to the other prediction problems, you can use:

- Logistic regression, decision trees or neural nets

How to do it:

Data preparation: Even if a data warehouse is accessible for the data miner, before you start your main analysis, try to find out whether the contract or service has a specific client article number or not. In many systems, you may be faced with the fact that one contract appears in several data cases because the contract contains several product elements each with its own article number and data footprint. Without this generated pre-knowledge, one customer contract can look like three contracts in the data.

Most of the time, the number of cases with Target = 1 (churn) is small. To construct a suitable dataset that can be used for model building requires a stratification strategy.

Business issues: Firstly, as discussed earlier, take care with the internal definition of churn. It will be different from company to company. Secondly, the target definition you decide upon should be discussed with colleagues responsible for working out the strategy to avoid churn. You should note that in some cases it is easier to predict churn than to find a suitable strategy to avoid it. The prediction is only useful for the business if there are strategies in place to avoid churn having been alerted to its likelihood. Work on developing these strategies can utilise the pre-knowledge gained by the close scrutiny of the data taking place during the modelling process. For example, if advertising pressure is identified as a churn indicator, then a strategy can be worked out that will ensure that less advertising is sent to those groups particularly sensitive to it.

Transformation: The purpose of transformation is to make the final models more robust. Sometimes, strange things can happen causing oddities in the dataset, and the model needs to be robust to all these artefacts. For example, a mistake could occur at any stage in the data input or transfer, or a change in customer behaviour may be noted but not recorded properly or a comment may be added without further adjustment, or some people may end up with a quitting reason even though they did not quit. The model needs to reflect the underlying truth in the data rather than any erroneous quirks introduced during the data input and manipulation processes. For this, we need to replace missing values, exclude outliers (or ameliorate them) and smooth the data. We also need to Normalise to use methods requiring Normality and standardise to reduce the effects of differing means and variances. Apart from these more statistical aspects, transformations are also necessary to create combined variables that summarise the information in the data. From one contract, you may create several variables in the following ways:

1. Calculate a variable for each service or product that says how long the customer accessed the product, in days, before the service stopped, or if it has not stopped, calculate the time until the end of the input period.
2. Calculate a variable for each service or product that says how many days have elapsed since the product or service was last ordered.

3. Calculate a variable for each service or product that says how many days have elapsed since it was stopped. When you calculate this variable, be careful about those people who did not quit and those people who never ordered the product. You have to think how these values will appear in the analytics.

These are just examples, and there are plenty of other combinations that could usefully be expressed as variables. Some are more complicated, for example, the time elapsed before quitting a first contract if a second one is accepted.

Transformation is very important to improve results; otherwise, models can be over-affected by vagaries in the data. Transformation is equally important where necessary for the newly calculated variables arising from the particular business requirements.

Analytics:

Partitioning the data: It has to be decided whether to split the data into training and test samples. Generally, this is the preferred option if there is enough data. If the number of cases available to 'learn' the model is quite small, then cross-validation might be better than data partitioning.

Pre-analytics: Pre-analytics will help give you more ideas for artificial or combined variables and to get a first clue which variables might have a relationship with the target. During this stage, it is also possible to gather important knowledge that might help the business.

Model building: We recommend using a decision tree. It has the advantage that some of the rules found might guide to potential solutions for how to avoid churn. If decision tree analysis is not available in your toolbox, you can use logistic regression or neural networks as well.

Evaluation and validation: Apart from the statistical numbers to ensure the model quality, you have to be careful that the model is reasonable in a business sense and that it can be transferred to future applications. The threshold predicted probability of churn is likely to be an issue of discussion. In general, a threshold of 0.8 (if the prediction is given on a scale 0 to 1) is a good starting point to search for the optimal threshold. Another possibility is to look for the predicted probability at the point where the cumulative lift is 2. Recall that a lift curve shows the proportion of Target = 1 (churn) customers for each predicted probability of churn divided by the actual proportion of churners in the dataset. The predicted probability of churn (on the horizontal axis) at the point on the lift chart vertical axis where customers are twice as likely to churn could be a good value to use as the threshold.

Implementation: In some companies, it is very useful to implement the model so that the results can be used at the service desk. If this is intended, then you need to make sure that the rules/models and the newly created variables can be calculated, for example, in an ERP system. Another way of implementing the models is just by calculating/applying the model regularly in the analytical environment and selecting the relevant customers for further treatment.

Hints and tips: It might be useful to discuss different treatments for different predictions.

How to sell to management: In most companies, keeping hold of long-term contracts is seen as a top priority, and so, selling the modelling and the results to management is not a problem. However, putting strategies in place to avoid churn might become a problem.

11.2 RECIPE 21: INDIRECT CHURN BASED ON A DISCONTINUED CONTRACT

The business problem behind this recipe is similar to that in Recipe 20, but the business process is different. In some countries and companies, contracts with no pre-defined end date are unusual. To clarify the issue, consider the following overview of possible variations:

1. Contract with open or undefined end.
2. Contract with a defined minimum end, for example, it can only be stopped by the customer after 24 months, but if the customer does not quit by a specific date, the contract will continue for another period, for example, for another year.
3. Contract will end at a specific date if the customer does not extend it.

Churn with contracts of types 1 and 2 can be predicted.

Churn with contracts of type 3 can be seen as a buying problem because churn happens if the customer does not buy the service/product again (after using it before). In this case, the churn problem turns into a problem that can be solved as described in the recipes for buying affinity in Chapter 8.

Target variable: The target is quite easy, being Target = 1 for all those with an old contract who ordered again in a defined time slot and Target = 0 for those with an old contract who did not order again.

Data preparation: It is worthwhile being creative in the choice of variables including the following data if they are available:

Input data – must-haves:

- Order and contract data (existing and former contracts or bought products)
- Data on received invoices
- Customer service/online services (like 'my account')
- Data on advertising activities (online and offline including newsletters)

Input data – nice to have:

- Payment data
- Usage history of services, software or knowledge

If we look back to the business problem, it might be an opportunity to do a second prediction for every customer with low affinity to reorder the same product or service as they might have a higher predicted affinity for another product.

11.3   RECIPE 22: SOCIAL MEDIA TARGET GROUP DESCRIPTIONS

Industry: The recipe is relevant to everybody with a strong digital marketing focus who needs to learn more about their digital target group.

Areas of interest: The recipe is relevant to marketing, sales and online promotions.

Challenge: The challenge is to learn more about the consumers who are related to the company-owned pages or accounts in social media. Based on these descriptions, the marketing strategies might be optimised.

Typical application: A typical application is defining target groups based on their social media data and behaviour. There are 3 ways to gather this information:

1. Use analytics services as provided by the social network itself or by third-party companies.
   a. Advantage: There is fast access because the results are delivered on a summarised level and there is no need for permission to access the data.

    b.   Disadvantage: Data is not accessible on an individual level, neither can you obtain different groupings on a similar level of aggregation to the one provided by the service.
2.   Use static grouping based on the data you have permission to use.
    a.   Advantage: You can define the target group as similar to the definitions used by the marketing department based on the data you have access to.
    b.   Disadvantage: New and so far unknown combinations might be overlooked. You have to have the permission to use the personal data and the behaviour data.
3.   Use analytical based grouping. This way uses clustering or Self-Organised Maps (SOM) to find homogeneous groups in the data you have permission to use.
    a.   Advantage: You may find unexpected groupings that give you new insight.
    b.   Disadvantage: You have to have the permission to use the personal data and the behaviour data. The new groupings may totally mismatch with all existing target group definitions.

In this social media recipe, we will concentrate on the first and second ways to gather information because the third way is a variation of Recipe 13 (clustering) and the only new aspect is how the data is used which will be described here. So, we just focus on social media-generated data and the use of given static groupings (see Figure 11.2).

Necessary data: This includes data collected when social media is used. Note that data protection rules differ from country to country. Also note that different kinds of social networks collect different data and give different data access to users compared to access given to the owners of company accounts or pages. We recommend that you consider this seriously and if in doubt ask the actual users for their permission. For example, if a user chooses to download a game or app or to use another service, then they should be asked to give their permission for their data to be used both for analytics and for further communication.

Population: The population is everybody registered or linked to a company-owned social media page or brand account in a social network.

Target variable: No target variable needed.

Input data – must-haves: This includes personal data and behaviour data (amount of post, followers, people to follow, activity on social media, etc.). The data are available through the Application Programming Interfaces (APIs) with the relevant permission.

FIGURE 11.2   Typical application.

Input data – nice to haves: A Single Sign-On (SSO) allows the user to log into a number of different systems simultaneously. If an SSO is available, then additional data for other networks might be accessible.

Possible data mining methods: The methods used are descriptive and exploratory statistical analysis.

How to do it:

Data preparation: The specific task in this recipe is the data preparation. The data out of social networks is sometimes unstructured or may follow a specific structure such as that in a JSON (JavaScript Object Notation) file. An example of a JSON file is the following:

{"users":[{"uid":"42830256","birthday":"17.06.1995","education":"university","logo50":"http:\/\/logo.XXXX1.com.cn\/logo\/83\/2\/50_42830256_2.jpg","trainwith":"","city":"london","online":"0","favtv":"","interest":"","name":"miller","isStar":"0","gender":"0","idol":"","favmovie":"","hometown":"leds","favbook":"","status":"0","career":"manager","logo120":"http:\/\/logo.XXXXX1.com.cn\/logo\/83\/2\/120_42830256_2.jpg","marriage":"0","intro":"","wishlist":"","motto":""}]}

As you may have noticed, the distinctive feature of the structure is that only the available information is stored. In this structure, the variable name is given first and the value comes second, for example:

"birthday":"17.06.1995"
"education":"university
"career":"manager

You need programming skills or the help of software to transfer this data structure into a format that is suitable for further analytics, for example, see Figure 11.3.

The format of the original file extracted out of the social network API may be different from one network to another. So, if you do not have strong programming skills, you will need to ask the IT department for help.

Business issues: The aim is to learn more about your social media target groups, and you can use the given information to describe them better.

Transformation: After the unstructured data is put into shape for analytics, you still need to consider making transformations for good results. For example, if the data has been typed in by hand, it is likely that there will be 'typos' and these must be corrected. The variable 'age' has to be calculated and may be classified into categories. If gender is given as a code and data is assembled from several social networks, make sure that the code is the same in every network!

Analytics: Analysis of the static groups proceeds in the same way as in the experience-based approach of segmentation described in Recipe 13. To carry out this approach, first, set up the rules for the experience-based segments, for example, one segment may be women aged 18–34 years old; then, group people in the different segments with all their associated variables.

An example of potential segments is shown in Figure 11.4.

In this approach, we calculate frequencies, ranks and means to help to describe the segments and their behaviour in detail.

uid	Birthday	Education	Trainwith	City	Online
42830256	17.06.1995	University		London	0

FIGURE 11.3   Typical dataset.

Segment	Age group	Gender
S1	0–17 years	Female
S2	18–34 years	Female
S3	35–54 years	Female
S4	55–74 years	Female
S5	Older than 75	Female
S6	0–17 years	Male
S7	18–34 years	Male
S8	35–54 years	Male
S9	55–74 years	Male
S10	Older than 75	Male

FIGURE 11.4   Typical segmentation.

Evaluation and validation: This recipe involves an unsupervised approach, and so, it is recommended to cross-check whether the results found fit in with the known picture of the target groups as learnt, for example, from market research. If they do not fit, then try to find out why.

Implementation and more: The implementation is as described in Recipe 13.

11.4   RECIPE 23: WEB MONITORING

Industry: The recipe is relevant to all industries, mainly those with strong customer relationships, high recommendations or owning great brands. Actually, it is most relevant to industries in business-to-consumer business.

Areas of interest: The recipe is relevant to marketing, sales and online promotions.

Challenge: The challenge is to find and learn about all bits of information, data and whispers regarding your products, service, brand, etc. Only the knowledge of those ongoing communications in the web will enable the company to interact in time (see Figure 11.5).

Necessary data: Web monitoring and especially text mining use words and phrases instead of numbers.

First of all, the whole web can be seen as a potential source of data. There are some public and freely available tools that claim to be able to find everything: A typical tool, for example, is 'Google Alerts'. This tool lists every piece of news or notes every time a specified keyword or key phrase shows up. Depending on the keyword, you will end up with quite a big unstructured list of sometimes hundreds of links that may or may not contain some relevant information for you.

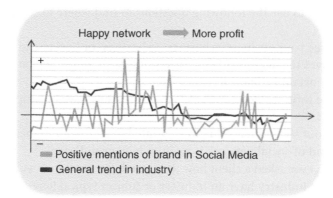

FIGURE 11.5   Typical application of web monitoring.

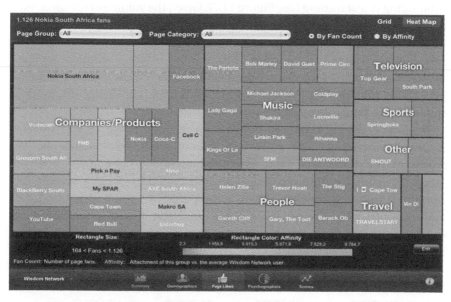

FIGURE 11.6   A potential result.

Generally, it is much more effective to put time into distinguishing between potentially relevant sources and potentially irrelevant sources than just checking out everything on the list so that you can go to the potentially relevant sources to find what you are looking for (see Figure 11.6).

We propose to use the following questions to find where to search first:

- Do you see your business as business to business or business to consumer?
- Who drives the decisions from the client side? A single person or a group?
- How important is image in your business?
- Are there forums, blogs or portals that you know have big impact on your business?
- What about societies, associations, printed magazines and their websites?
- What kind of social network is relevant to your industry?
- Have you ever asked a client how they search for news from your industry?
- Does your industry have an industry or subject typical language and what kind of words would a naive person use to search for the same subject? Would both groups use the same words to find the same information?

Possible data mining methods: The most common technique to do web monitoring is text mining (see Figure 11.7). Given the volume of text generated by business, academic and social activities – in, for example, competitor reports, research publications or customer opinions on social networking sites – text mining is highly important. Text mining offers a solution to these problems, drawing on techniques from information retrieval, natural language processing, information extraction and data mining/knowledge discovery as the following figure illustrates.

In essence, during enhanced information retrieval (stage 1), sophisticated keyword searches retrieve potentially relevant electronic documents. The words of the document (and associated meta-data) are then processed (stage 2), using, for example, lexical analysis (aided by domain-specific dictionaries), into a form that allows a computer to extract structured data (information) from the original unstructured text. Useful information can then be extracted from the documents (stage 3). The identified information can then be mined to find new knowledge and meaningful patterns across the retrieved documents (stage 4) which would be difficult, if not impossible, to

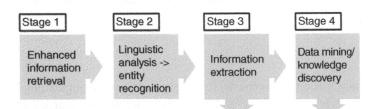

FIGURE 11.7 Text mining.

identify without the aid of computers. Exactly how and what can be achieved depend on the licensing, format and location of the text to be mined.

Text mining is the discovery of previously unknown information or concepts from text files by automatically extracting information from several written resources using computer software. In text mining, the files mined are text files which can be of one of two forms:

1. Structured text
2. Unstructured text

Structured text consists of text that is organised usually within spreadsheets, whereas unstructured text is usually in the form of summaries and user reviews. Unstructured data exists in two main forms: bitmap objects and textual objects. Bitmap objects are non-language based (e.g. image, audio or video files), whereas textual objects are 'based on written or printed language' and predominantly include text documents.

How to do it:

Data preparation: You need to find the right places to search for information. Use our checklist in the preceding text to help set up the keyword and key phrase that reflects your business.

The usual data mining processes for predictive modelling are not suitable for text mining, so descriptive analytics and data partitioning are not necessary.

Model building: Association rules, sequence analysis and decision trees are the most commonly used families of algorithms. Apart from methods that have their roots in data mining, there are also other algorithms that have been developed, including the following:

Token scanner: Text structure (paragraphs, titles, characters, paragraphs, etc.) and special strings (punctuation, dates, acronyms, HTML tags, etc.) are identified.

Lexical analysis: Morphological analysis of word forms (definition of the word class with part-of-speech tagger) and the inflected form (plural/singular, etc.).

Named entity recognition: Finding special words such as person, company, product, place names and complex dates, times and measurements.

Parsing: Highly modularised and thus allows easy but domain-independent analysis with phrases and very domain-specific rules for the recognition of complex units.

Core reference identity: The central task is to determine whether different linguistic objects refer to the same template instance reference. This includes determining whether different words describe the same thing or person, for example, 'President Bush', 'George W. Bush' and 'Bush', and the resolution of pronominal reference (references between pronouns, proper names, phrases) and references between designators ('the software giant') and other entities ('Microsoft').

Domain-relevant pattern recognition: Finding characteristics of the heads of the extracted phrases filled with the individual characteristics of a domain-specific template instance.

Template unification: As the information sought may be spread over several sentences and paragraphs, it is necessary to combine information from different template instances. In general, this complex task is established by unifying two templates that have at least a similar attribute which could be that they include similar words or are part of a relationship.

The latter two components require domain-specific knowledge and are very time consuming in their preparation. They require expertise and customisation and are generally automated using methods of machine learning.

Evaluation and validation: Apart from looking at the statistical numbers to ensure the model quality, you have to be careful that the model is reasonable in a business sense.

Implementation: Web monitoring results are mainly used to improve communication and public relations strategies, and they particularly influence strategies regarding the social web. Web monitoring results are not deployed in the same way as the results of prediction modelling, but they provide general background knowledge.

11.5   RECIPE 24: TO PREDICT WHO IS LIKELY TO CLICK ON A SPECIAL BANNER

This problem can be solved as a predictive model and is a variation of Recipe 1. Clicking on a specific banner is similar to reacting to a specific marketing campaign. But there are some issues that need special consideration. These issues are discussed later.

Lack of information: For most banners, there is no information available as to whether they are viewed by any user or not. For a typical direct marketing campaign, this information is usually available and can be used to limit the

ENBIS_Challenge_2...
Quelle

Spalten (938/0)
userid
x1
x2
x3
x4
x5
x6
x7
x8
x9
x10
x11
x12
x13
x14
x15
x16

Zeilen
Alle Zeilen 73.730
Ausgewählt 0
Ausgeschlossen 0
Ausgeblendet 0
Beschriftet 0

userid	8	x9	x10	x11	x12	x13	x14	x15	x16	x17	x18	x19	x20	x21	x22	x23	x1000	x1001
55448	3	3	0	0	0	0	0	0	0	0	0	0	24	0	3	3	24	
55449	10	20	0	0	0	0	0	0	0	0	0	0	493	30	50	99	71	
55450	10	37	0	0	0	0	0	0	0	0	0	0	4319	0	145	626	193	
55451	2	2	0	0	0	0	0	0	0	0	0	0	3	0	2	2	3	
55452	10	13	0	0	0	0	0	0	0	0	0	0	44	0	11	14	42	
55453	1	3	0	0	0	0	0	0	0	0	0	0	28	0	1	3	28	
55454	10	14	0	0	0	0	0	0	0	0	0	0	5018	0	50	70	502	
55455	10	19	0	0	0	0	0	0	0	0	0	0	3950	0	88	142	161	
55456	2	6	0	0	0	0	0	0	0	0	0	0	70	0	2	6	17	
55457	10	17	0	0	0	0	0	0	0	0	0	0	254	0	18	31	111	
55458	10	10	0	0	0	0	0	0	0	0	0	0	425	7	86	97	36	
55459	10	11	0	0	0	0	0	0	0	0	0	0	3473	0	58	59	503	
55460	4	3	0	0	0	0	0	0	0	0	0	0	56	0	4	3	54	
55461	1	1	0	0	0	0	0	0	0	0	0	0	1	0	1	1	0	
55462	10	18	0	0	0	0	0	0	0	0	0	0	1271	4	62	156	28	
55463	10	14	0	0	0	0	0	0	0	0	0	0	52	0	13	18	40	
55464	10	19	0	0	0	0	0	0	0	0	0	0	306	0	14	25	168	
55465	4	4	0	0	0	0	0	0	0	0	0	0	38	0	4	4	36	
55466	6	6	0	0	0	0	0	0	0	0	0	0	89	0	6	6	89	
55467	4	7	0	0	0	0	0	0	0	0	0	0	54	0	4	7	53	
55468	2	4	0	0	0	0	0	0	0	0	0	0	22	0	2	4	22	
55469	10	13	0	0	0	0	0	0	0	0	0	0	391	0	35	42	107	
55470	2	5	0	0	0	0	0	0	0	0	0	0	31	0	2	5	31	
55471	10	17	0	0	0	0	0	0	0	0	0	0	196	0	14	21	140	
55472	9	11	0	0	0	0	0	0	0	0	0	0	44	0	9	11	43	
55473	8	8	0	0	0	0	0	0	0	0	0	0	11	0	8	8	10	
55474	7	36	0	0	0	0	0	0	0	0	0	0	110	0	7	36	100	

FIGURE 11.8 ENBIS Challenge DATA.

population to whom the marketing is directed. This is not normally possible for banners published on websites.

Disproportionate reaction rates: If we consider that a typical banner will reach a couple of million people during the time it is published but only a few hundred will click on it, we are faced with a significant mismatch between those who reacted (Target = 1) and those who did not react (Target = 0). If we compare an offline campaign such as an ordinary mailshot with using a banner, we will see a marked difference as illustrated in the following:

Mailshot (offline): Mailshot is sent to 500 000 prospects and 3 000 react giving 0.6% reaction.

Banner (online): Banner is visible to 5 000 000 people and 800 click on it giving 0.02% reaction.

Analysing the reaction to a banner requires sophisticated sampling. We recommend creating an artificial sample by taking several samples out of the 800. This could be done by taking a random sample of 500 out of the 800 and repeating this 10 times to get a dataset of 5000 for Target = 1. For the Target = 0, we recommend to take a sample of 15 000 out of the 5 000 000 people who may have seen the banner. The dataset for analysis is therefore stratified 1:3.

Shape and content of data: The shape and content of the input data is quite different. You should check how the past clicking and web usage data of users and visitors to the website is stored. It is possible that the data is online and available as a log file in which case the main focus is on creating a suitable dataset. You should end up with a dataset that is quite similar to the example dataset in the ENBIS Challenge as shown in Figure 11.8.

To generate data in this sort of format, just think in general terms because it is unlikely that your variables can be transferred to future usage. This is especially true when dealing with the web as it changes so quickly. For example, consider the following situation:

> You count clicks on a sub-page that contains information on a single event such as 'election to the Bundestag 2013'. Provided you map the information to more general variables related to politics, elections, national politics, etc., then it is easy to reuse the resulting model in future, but if you sum up the clicks in a variable called 'elections to the Bundestag 2013', you cannot reuse the model for prediction because this event will never come back. Such a model can only be used for explanation.

Last thing to consider: The model you worked out should be fast and easy to implement in the business environment. Sometimes, the model must be translated into 'if then else' rules so that a system like the 'Adserver' can use it to send the banner out to the right people.

12

Software and Tools: A Quick Guide

12.1 LIST OF REQUIREMENTS WHEN CHOOSING A DATA MINING TOOL

There are several data mining and statistical software products that may help you to do data mining. To choose the tools that fit your needs, it is necessary to do a small requirements study. The following questions may help to discover

A Practical Guide to Data Mining for Business and Industry, First Edition.
Andrea Ahlemeyer-Stubbe and Shirley Coleman.
© 2014 John Wiley & Sons, Ltd. Published 2014 by John Wiley & Sons, Ltd.
Companion website: www.wiley.com/go/data_mining

your specific needs. We know that some questions are not relevant for your specific situation, but they may help you to rethink. Please also keep your normal requirements for software in mind.

Besides the content and technical aspects, the financial aspects differ a lot. There are some quite good free products on the one hand and some very good, more advanced solution packages for buying or to be licensed. As with other software issues, the type of machine on which the data mining software will run also has an impact on the price.

In general, we can separate the tools into three groups:

- Specialised tools for carrying out one algorithm or method, for example, to allow you just to do association and sequence analyses or neural networks or decision trees.
- Data mining tools with a very strong focus on a small group of data mining methods and with very limited possibility to arrange, aggregate or transform the data.
- Full data mining suites that support the whole data mining process. Sometimes, these tools also provide the option to translate everything needed for regular application/deployment in a more widely used software language like C++, SQL or SAS. This gives the opportunity to automatically include the deployment and regular routines necessary to make the result available without the involvement of the data miner themselves.

A list of tools is given on our website.

The following questions may guide you to a clearer picture of your needs: please rate all questions according to your situation from 'not relevant at all' to 'must-have'.

Topic 1: Economic advantage of the tools:

1. What is the economic advantage created by the new data mining tool?
2. How well does the data mining tool fit the industry-specific conditions?
3. What additional information can be obtained by the tool?

Topic 2: Technical knowledge:

1. What technical skills are required for users?
2. Can the departments depend on being able to work with the tool (without regularly calling for the IT department or the database specialists)?
3. Is the tool largely self-explanatory?

Topic 3: Comprehensibility and interpretability:

1. Are the analysis results intuitively understandable?
2. Is there a clear explanation of the analysis results?
3. Are the guidance notes in a mathematical and engineering form or in a common language?

Topic 4: Independence of the department:

1. Is it necessary to go to an analyst to address a more in-depth question?
2. What kind of follow-up questions can the departments answer with the tool?
3. How easily can you ask further questions during the current analysis?

Topic 5: The users:

1. What technical/statistical requirements does the user need to bring?
2. For how many users is the tool designed?
3. Are there different levels of users (department/analysts)?
4. Can the users develop their own applications/questions for the tool?

Topic 6: Reliability of results:

1. How reliable are the results?
2. Is it likely that 100 users will receive 100 different answers to the same question?
3. Are there correlations that cannot be discovered by using the tool?

Topic 7: Automation of analysis:

1. Does the tool have the ability automatically to create standard analyses?
2. Can the tool automate pre-defined data mining issues, such as retraining the predictive model in the first week of every month?

Topic 8: Data handling:

1. How much data can be processed by the tool?
2. Can the tool get data directly from the database or must the data be extracted beforehand, for example, from a Comma Separated Variables (csv) file?
3. If the tool works with samples from the data, how is it assured that no relationships and patterns in the data are lost?

Topic 9: Methods:

1. Are supervised and predictive methods available, such as decision trees and regressions?
2. Are unsupervised methods available, such as clustering and associations?
3. Are methods for feature reduction directly accessible by the analyst?
4. Are methods for descriptive and exploratory statistics implemented to explore the data?
5. Are there opportunities to do different sampling and to follow different sampling strategies by using the data mining tools?

Topic 10: Evaluation and validation:

1. Are validation methods like 'test and train' or 'leave one out' implemented?
2. Are the results presented graphically?
3. For classification methods, are there opportunities to define your own thresholds?

Topic 11: Deployment:

1. Is it possible to deploy the model on new datasets (of any size) through the data mining tool?
2. Is the time needed for deployment reasonable and does it match with your business needs?
3. Does the deployment automatically include all necessary data manipulation?
4. Are there options to convert the complete code needed for deployment to a more common software language (like C++, SQL or SAS)?

Topic 12: Integration with existing applications and IT landscape:

1. How does the tool fit into existing applications and IT landscapes?
2. How does the tool interface with other systems?
3. Can the analysis results be transferred for use in other tools (e.g. Excel)?

Topic 13: Support and training:

1. What support is needed for the installation and implementation?
2. Does the manufacturer give support and training?
3. Is there a hotline to the manufacturer for future reference?

12.2 INTRODUCTION TO THE IDEA OF FULLY AUTOMATED MODELLING (FAM)

To detect fast changes in customer behaviour or to react in as focused a manner as possible, predictive modelling must be done well to get effective predictions of customer behaviour, and it has to be done fast to be relevant for the business. Modelling speed is of great importance in industry as time is a crucial factor. This necessity requires a different technical setup for model development to fulfil both needs: quality and development speed. Today, most companies like to develop their models individually with the help of specialists. But for a lot of companies, this takes far too long; even though the models are excellent, the time to develop them sometimes eliminates the advantages of a better prediction. In the following, we describe the general structure and ideas on how to implement industry-focused model production that will help to react quickly to changing behaviour. We will discuss the key success factors and the pitfalls of this assembly line model product.

12.2.1 Predictive Behavioural Targeting

For a company to receive exciting customer profiles, improve the relevance of its online offerings and optimise its long-term online marketing ROI, it would not only need information about the historical and current habits of its customers but also about their future activities. So, it is important to discover patterns in customer behaviour, and for this, we need to identify a specific user. This makes the predictive modelling behavioural targeted.

Methods such as descriptive statistics, clickstream analysis, discriminant analysis, regression methods, decision trees, neural networks, cluster analysis and time series analysis are used.

Based on the analysis of user profiles and user structures (such as age, lifestyle, peer group affiliations, browsing behaviour), predictive models are created for future behaviour. For example, the decision on where to place advertising banners that customers will be shown is based on the sites visited or on the basis of what was done on these sites.

In previous and current times, contextual advertising, such as banner advertisements for suitcases, cheap flights and hotels, is linked to the content of special websites, in this case around travelling. So, the actual and past behaviour of the user/cookie was not relevant for the banner even if there is a clear logical link, for example, that the person interested in travelling has travel needs. Now, we work out predictive models based on the user's actual and past behaviour on the web to predict whether the person has a special need for a product, and then, this product will show up as a new banner. Predictive behavioural targeting based on who is the right person opens the way for relevant advertising. This clearly expands the opportunities for advertising.

Predictive targeting is immensely valuable when information about the user is available in the right form as quickly as possible. Modelling real-time online behaviour creates the conditions for fully automatic predictive targeting.

12.2.2 Fully Automatic Predictive Targeting and Modelling Real-Time Online Behaviour

Using the necessary algorithms for analysis in real time offers a head start and makes it feasible to use the new fully automated predictive targeting with its individual and lasting forms of communication.

Automated predictive targeting is an evolutionary step change comparable to the change from handicraft to manufacturing production. The classical way to build complex forecast models by hand is like the role of 'Master Workshop', but even a large staff of analysts (craftsmen) cannot cope with the huge amounts of data now available and the high number of models that can be made – this needs a fully automated 'assembly line' for the efficient production of prediction models.

12.3 FAM FUNCTION

The core of fully automatic predictive targeting systems is the construction of prediction models. It includes all functionality for a team of analysts to build complex predictive models by hand ('Master Workshop'), but also in a second module ('assembly line'), it builds very fast fully automated simple, click-based predictive models, automatically backs up its quality and makes it available for use.

In the 'assembly line', all models are calculated, which is a relatively simple task in the field of predictions (by predictive modelling), because, for example, we only build models whose target variable is a dichotomous structure (such as clicked vs. not clicked, purchased vs. not purchased, visited vs. not visited). These prediction models cover, for example, a large portion of the orders for banners and optimisation behaviour targeting. Special analyses such as cluster analysis are performed by the analysis team in the 'Master Workshop'.

It is decided by an administrative process whether a predictive model goes into the workshop or into the assembly line to be manufactured. Each model receives a clear ID and is archived. The substantial elements or modules contained in the assembly line are shown in Figure 12.1.

Modules of the automatic targeting system

- **Control**
  Administration
  Model management
  Management of the variables

- **Learning of the model**
  Selection of the target variable = 1 and target variable = 0
  Model specific random sampling to obtain a dataset
  Supply of a learning dataset and a test dataset
  Learning the model
  Evaluation of the model
  Quality assurance monitor
  Release process
  Handing over process

- **Model examination**
  Model specific random sampling to obtain a dataset
  Model examination
  Quality assurance monitor
  Confirmation process

- **Model archive**

- **Variable processing and supply**

FIGURE 12.1   The elements of an assembly line.

12.4   FAM ARCHITECTURE

The areas of 'Master Workshop' (work carried out by the analysis team) and 'assembly line' (automatic targeting) are included in the existing architecture of a company in a way that the environment and its benefits can be used as far as possible including tasks like data preparation.

In general, all the developed models will be passed as a code/script and archived in an archive, including their meta-data and also details about their use. The application of models based on the scripts is the final step of the calculation of variables at the end of a session or a slot. For each active model, the prediction is calculated as a relevant forecast value per Unique Client (UC) and is stored in a separate variable and made available to the Target Builder.

In the Target Builder (an instrument for targeted delivery of content), these predictive capabilities of profiles are used to provide target audiences for

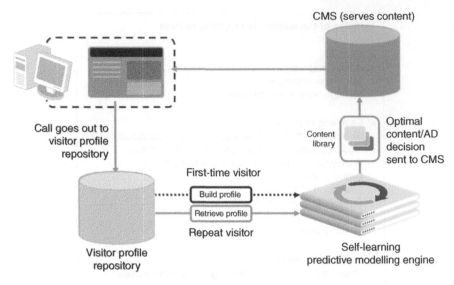

CMS (serves content)

Call goes out to
visitor profile
repository

Content
library

Optimal
content/AD
decision
sent to CMS

First-time visitor

Build profile

Retrieve profile

Repeat visitor

Visitor profile
repository

Self-learning
predictive modelling engine

FIGURE 12.2 Automatic predictive targeting.

online campaigns and make them ready to be sent to marketing. So, every user with a suitable profile is addressed (see Figure 12.2).

12.5   FAM DATA FLOWS AND DATABASES

The databases for all analyses are the behaviour data of the UC on the respective websites. Ideally, the behaviour data can be enriched by information from questioning or login data. One can differentiate between standard/general behaviour and interest-conditioned information. These profile and behaviour data per UC are computed relative to different time windows. Both are formed from large numbers of variables. The variables should reflect different views on the UC.

In the context of the modelling in the assembly line, only an examined subset of variables is used, in order to grant stability, robustness and performance essential for automation. That can be seen somehow as the standard dataset. The size and content of it is the outcome of experience and domain knowledge obtainable from the analyses team. Using the Pareto principle, the dataset should contain roughly 20% of all potential data to be the basis of 80% of all easy prediction problems. To ensure that data preparation fits the time constraints of the assembly line, it is important to use optimisation techniques that help to deliver the data as fast as possible. After the structure of standard dataset is defined and adjusted, it will be created fully automatically by the systems.

As in other industrial processes, it is important to identify and separate different process steps, such as data handling and calculations so that they can be standardised and optimised to deliver an outcome of reliable quality and with a more or less small standard error. About the data preparation including all necessary (automatically done) transformations, it leads us to the fact we should choose robust methods that will help to generate data that fits most (modelling) situations but might not be the optimal one for every single case as you would get, when you do the analytics by hand through an expert in your company. Where you try to optimise the modelling result by a cycle of data preparation, analytics and measurement, especially under time constraints, the number of cycles has to be at a minimum.

An essential part of the daily profile building step is data preparation. It is needed both for the modelling/verification and for the application of existing scores. All of this has to be done in a time-critical area.

So that modelling or deployment can take place in the current session, it is indispensable that the session data can be accessed at any time in the session.

To train and validate a model, very fast sampling is mandatory. Please keep in mind that the relation between those who act (Target = 1) and those who do not act (Target = 0) is very unbalanced. It is very likely that you have just 500 UC acting and 5 million UC not acting. As part of the modelling issue, it makes a big difference in calculation time whether you have to calculate slightly more than 5 million records or just a couple of hundred. So, if you start to build an assembly line, it is necessary to work out and test a sampling strategy that best suits your individual situation.

12.6   FAM Modelling Aspects

Similar to the data preparation, the assembly line needs for its modelling engine a data mining algorithm that will deliver good and stable models without any interaction with an expert. The algorithm should be fast, and the time needed for the deployment of models should be as short as possible, especially if it is planned to use the assembly line for nearly real-time forecasting. The time for modelling and deployment might not be so time critical if for business reasons it is not so important to be in 'real' time. For example, it is sometimes sufficient to use data from the last finished session as the newest input to the forecasting, or the forecast may be shaped by the content last clicked on or searched for. Based on many tests simulating the situation like an assembly line, out of all the algorithms, the family of decision trees wins mostly and speedily delivers very good results.

The fully automated quality control is the next step in the process to deal with; the task here is to define the boundary between not 'quite as good' and 'good enough for the business'. If your expectations for the modelling quality are too high, you will end up with lots of models transferred back from assembly line to workshop to be redone by the experts, so you will lose time and it will cost you more money to produce the models. If your quality is too low, you will lose business, for example, through lower click rates.

In quality control, we are looking after new (freshly developed) models as well as after models that have been in duty for a while, so the chosen quality control process and its measurements must be able to do a constant quality control on all active models to notice in time when a model needs refreshment.

12.7   FAM Challenges and Critical Success Factors

The complexity of the process of fully automated predictive targeting and modelling of real-time online behaviour presents some statistical challenges, for example when a model is built using historical data then sample size is usually not a problem as we have a lot of data and know in advance the size of the dataset; however, when we are building a model automatically, then we want to start as soon as possible, but we do not have sufficient information to decide how big a sample is needed nor how to handle the stratification to be able to build a good model. Note that the target variable changes from 0 to 1 whether a person clicks on the item of interest. Forecasting methods are judged in terms of their prediction quality, stability and longevity. Performance, durability, runtime behaviour, parameterisation and automation of error detection must be considered when selecting the quality assurance methods.

Similarly, critical success factors should not be ignored: Are uncontrollable optimisation steps/algorithms influencing the advertisement server that originally predicted massive click behaviour? Are tags missing in the banner? Are there so few clicks that it takes too long to get a critical mass for modelling?

12.8   FAM Summary

Fully automatic targeting and predictive modelling of real-time online behaviour are at the very beginning of their development and are valuable tools provided critical success factors and requirements are carefully observed in the implementation and the environment. All being well, we can obtain fully automated predictive targeting even based literally on the last click in real time, and these predictions can be flexible and up to date. This allows rapid response to changes and trends in the rapidly changing online marketplace.

13

Overviews

Overviews

A Practical Guide to Data Mining for Business and Industry, First Edition.
Andrea Ahlemeyer-Stubbe and Shirley Coleman.
© 2014 John Wiley & Sons, Ltd. Published 2014 by John Wiley & Sons, Ltd.
Companion website: www.wiley.com/go/data_mining

13.1 TO MAKE USE OF OFFICIAL STATISTICS

Official statistics are gathered by National Statistics Institutes and are a free resource for businesses and data miners. The quality of official statistics is assured by regulated checking methods, and historic series of matching data are available. Compatibility across countries is also generally good. Official statistics can inform a business about regional hotspots of potential customers, opportunities for specific types of products and services and neighbourhood networks. For example, official statistics are available for population size stratified by age, gender and small geographic areas.

The wealth of information available is remarkable. It is possible to find data on health, employment and education as well as composite data such as distance travelled to work and population flows in and out of areas. If continuous variables are transformed by categorising or binning, then the bins should align with official statistics where possible, for example, age and education measures. It is then possible to use base figures from official statistics to establish the potential impact of your modelling. If your analysis predicts increasing sales in a certain demographic group, you can check which areas are likely to be best for your promotional activities or where to target a new business venture.

Official statistics are the third source of information and knowledge available to the data analyst: firstly, there are business considerations; secondly, there are the results of analytics; and thirdly, there are the insights obtained from other research knowledge including that available in official statistics.

13.2 HOW TO USE SIMPLE MATHS TO MAKE AN IMPRESSION

Figures are extremely powerful as a way of communicating but only if they are delivered in a meaningful way. Here, we consider some tips.

13.2.1 Approximations

Even if you cannot find exact figures, for example, when calculating the benefit of some work or an applied technique, you can still give some idea of a possible range and give upper and lower bounds. We may not always know the full details of all costs; however, we may know an approximate value, for example, hourly rates may range from minimum wage up to the most generous amounts. Substituting a good guess or an upper or lower value will enable you to communicate an impression much more strongly. For example, suppose your data mining model saves an hour a week of meeting time, you can easily calculate the value of the time saved and increase the impact of your work: note the number of people that usually attend the meeting and multiply by their hourly rate. If 10 people usually attend the meeting and you estimate an average salary of a very optimistic £52 000 pa, then each week costs £1 000 and each hour costs £1 000/40 (assuming a 40 hour week) plus 20% employment costs giving £30 per hour. Your data mining work then gives a saving of £300 each week or approximately £15 000 pa.

13.2.2 Absolute and Relative Values

A different picture is given by presenting results in absolute and in relative terms. For example, there could be an increase in weekly sales company-wide of £100 000, but this could give quite a different picture if presented as a relative value: an increase in sales per weekly throughput or per branch or per customer or per opening hour. Subdividing by types also gives a more useful result and the possibility of identifying areas of excellence and areas in need of improvement. For example, we can present the results in different units and granulation such as sales per day or per week or per year and mean sales per outlet, per campaign or per new customer.

13.2.3 % Change

We can also give year-on-year change or percentage change. It is always good to include the actual figures and to plot the data in an appropriate chart. If the denominator of a ratio is small, then small changes can produce big changes in value. For example, if there is only one customer in a particular region and then another is procured, then there is a doubling of customer numbers, but this may not really be so impressive. Similarly, if the lone customer is lost next year, then the change in customer numbers is −100% year-on-year and will show up as a major loss in the annual results.

13.2.4 Values in Context

A major consideration here is the quality of the information that is transmitted which will depend on the task that is being addressed. There are many cases where knowing the context of the figures avoids wasteful panics or adulation. For example, if the number of complaints to a department is 25 in one month, what should be done? There might be suggestions to (i) check the counting and (ii) check the department staff. However, a plot of complaints received every month by that department over the last year may show that numbers vary between 10 and 40 and that 25 is a typical value. This does not mean that no action should be taken, but it does mean that the problem has not changed significantly and can be tackled on that basis.

13.2.5   Confidence Intervals

The results produced from samples are estimates and depend on the sample used. When the sample is small, the variation between results from different samples can be considerable. In that case, it is important to give a confidence interval. However, the size of the confidence interval reduces with increasing sample size, and so, if the sample size is very large, then the confidence interval will be very small.

13.2.6   Rounding

Management and any more casual reader of results will prefer to see the results without the noise. This suggests that it is a good idea to round the results to whole numbers where possible and use a suitable number of decimal places.

13.2.7   Tables

Lists of numbers are seldom appealing. Tables can give a clearer picture, for example, pivot tables in Excel are very useful. In data mining, much use is made of summary tables such as the confusion table which aims to give an impression of the success of the modelling.

13.2.8   Figures

Data visualisation is an advancing field. IBM and Google have labs focusing on it as does the UK Office for National Statistics. Annotation on graphs adds the personal touch. Graphics are vitally important both to understanding data and to presenting results of analysis. Software offers an overwhelming choice of styles of graphics and multiple factors can be included in a graphic. The aim is to convey information without confusing the message. In particular, colours and shapes can be used to good effect.

The main plots encountered in data mining are line charts such as lift and gain charts, tree diagrams, histograms particularly as part of pre-analytics and scatterplots to illustrate important relationships.

13.3   DIFFERENCES BETWEEN STATISTICAL ANALYSIS AND DATA MINING

Data analytics includes statistical analysis and data mining. Although they have a lot in common, they are also different in some interesting ways. The following descriptions are drawn from the preceding text.

13.3.1   Assumptions

Statistical procedures usually require certain conditions to be met in the data, for example, there are often assumptions about Normality, linearity or constant variance. In contrast, however, in data mining, it is expected that these conditions are not met. Data mining procedures are usually carried out with very large amounts of data (random sample sizes of 30 000 are quite common), and the non-observance of the conditions does not matter too much.

13.3.2   Values Missing Because 'Nothing Happened'

Missing data that represents 'nothing happened' or 'information not stored' is very different to real missing data. These cases happen because most company databases just store things that happened and not things that have not happened. An example is that customers 'who buy' create footprints in the database; that fact is stored in several tables in which you will find the purchase date, amount, value, products, the way they are ordered, paid for, delivered and so on. For those customers 'who did not buy', you will find nothing in the company databases at that time. But, for some analyses, it is important to represent the 'not buying' fact in the dataset as well, especially if 'not doing something' is done by the majority of people.

Instead of representing the 'not happened case' by zero, you can also count the number of days since the last time something happened. For example, every 'no buy' can be represented by zero if the variable itself contains values (money) or amount (pieces), but if the variable contains 'days since', then you cannot represent the missing data with zero because that would wrongly lead you to the interpretation that something happened just recently. Here, it may be better to use the number of days since the last known activity (e.g. subscribe for an email newsletter) for estimation or other similar substitutes that correspond with your business rules.

13.3.3   Sample Sizes

A major consideration in statistical analysis is the size of the sample that is available. Often, respondents are few in number, and the lack of information is magnified if data are categorical or poorly distributed, for example, we might only know whether a person agrees or disagrees with a proposal or the distribution of amounts spent may be highly skewed. In these cases, statistical tests are not very powerful so that it is difficult to prove anything of interest conclusively. However, in data mining, there are usually more than enough customers (or cases) available for almost all questions in the analysis. Usually, sample size is not an issue.

Representative random samples are often extracted from the full dataset for the analysis not only because it speeds up the calculations for the modelling but also because we can then test and validate the models and be more confident that they are robust to changes in the population.

13.3.4   Goodness-of-Fit Tests

An overall measure of goodness of fit for regression models is conveniently summarised in the $\%R^2$ value. This is the % of the variance of the target variable which is explained by the model. In statistics, this value should be reasonably large for the analysis to be meaningful, for example, values more than 60% are considered as satisfactory. However, in data mining, particularly when estimating buying behaviour, these values are practically never reached. Here, a $\%R^2$ value as low as 11% might be considered very good. Even very low $\%R^2$ can indicate a useful model as at least there is some indication of a relationship between input and target even though confidence intervals may be wide.

Furthermore, according to experience, one should be very critical with a model for, say, buying behaviour, if the $\%R^2$ value reached is greater than 20%. The reason for this is that the noise in the data usually leads to low $\%R^2$ values, and so, if a value greater than 20% is reached, it is almost always the case that one or more of the input variables separates the target cleanly in the period of analysis, for example, if the input variable is purchase location = United Kingdom or Germany and the target variable is buy or no buy, then a large $\%R^2$ could arise because the product is only sold in the United Kingdom.

Also, there may be overlap between input and target, for example, one input variable could be a flag for a certain behaviour occurring, and it could happen that the behaviour took place at the same time as the target was achieved. For example, if a variable represents how many cars someone has bought but it is not time stamped, then it is very likely that the last car they bought is also included in the data summation. If the target variable is buy or no buy for a particular new car, then the number of cars bought is likely to be highly correlated with the target producing a model with a high $\%R^2$ value. The solution is to be more careful about subtracting the car of interest from the summation (and any other cars bought after this one).

To find candidate variables for this sort of problem, look at the importance ranking of each variable given by the data mining tool or the % of contribution. You should find the useful candidate within the first three variables, and it will have an enormous contribution to the model which might be suspicious and should be cross-checked with domain knowledge. There is still a chance that the model is working and is a good model, but it has to be checked. An easy method to check is to construct a contingency table for the target and the suspicious variable, and if the table has nearly 100% frequencies on the diagonal, then check the timeline very carefully. Note that if the model is being used for explanation or description, it is not really a problem and is in fact an interesting finding; however, if the model is being used for prediction, it is unlikely to work on new data.

13.3.5 Model Complexity

Models are valuable if they have good forecast ability and their mistake rate as indicated by the confusion matrix is low. In addition, it is important in practical data mining to make sure the models are plausible and useful. According to the type of problem being addressed, it can make sense to prefer models that use a large number of variables. This is contrary to the usual statistical practice of preferring models with a lower complexity (i.e. as few relevant variables in the model as possible). However, in the application of forecast models, it can make sense to obtain a very detailed ranking of the customers. This requires that each group of customers identified contains a reasonably small number of customers. This satisfactory situation is more likely to be obtained if the model contains many variables which will contribute to the segmentation of the customers by the model.

13.4 How to Use Data Mining in Different Industries

As discussed in Chapter 6, data mining methods can be divided into supervised and unsupervised methods. In addition, the usage of data mining can be divided into description and prediction.

Regardless of the industry where data mining is going to be used, the first thing to decide is whether the solution is needed to describe an existing situation or relationship or to predict a situation or result that may happen in the future.

Just to clarify, a description just describes what happened in the past; it is not necessary to control whether the input data are available in the future as well. Descriptions can be generated using supervised and unsupervised methods. Prediction, however, is always based on supervised methods. It is important that the same kind and quality of input data is available for training, testing and deployment.

The most common solution is a prediction model, as in Recipe 1. As already discussed, the key issues to consider are the definition of the target, the potential input data and the cut-off criteria. In the following list, we give a typical business question, a potential target variable and potential input variables for different industries. This list should help to map the solutions described for sales and marketing problems in the aforementioned recipes into other industries. The list is not complete but it gives a few ideas that might help to adapt a suitable solution.

A

Industry: Production

Question: Is it possible to predict the quality of a workpiece based on an image taken?

Target: Workpiece is fine (yes/no).

Input data: Data calculated out of single images of workpiece fine and workpiece defective. The data should contain shapes, green shadows, light/dark levels, etc.

Cut-off: The cut-off depends on the cost of a decision that turns out to be wrong. For example, the workpiece is predicted to be OK, but the prediction was wrong and the workpiece broke down. If the cost of the wrong decision is very high, it might be better to choose a cut point quite high.

Kind of problem: Prediction

Recipe numbers to look at: 1 and 2

B

Industry: Production

Question: Is it possible to predict the quality of a workpiece based on production data?

Target: Workpiece is fine (yes/no).

Input data: Sensor data and other data out of the production line. Note that the order of the data follows the production process and that some data might be correlated to each other.

Cut-off: The cut-off depends on the cost of a decision that turns out to be wrong. For example, the workpiece is predicted to be OK, but the prediction was wrong and the workpiece broke down. If the cost for the wrong decision is very high, it might be better to choose a cut point quite high.

Kind of problem: Prediction

Recipe numbers to look at: 1 and 2

C

Industry: Production

Question: Can we predict the optimal amount of material that will be needed during a particular time period?

Target: Number of items (of a single material) needed in a period

Input data: Order and sales data and data from other production areas

Cut-off: The cut-off depends on the cost of a decision that turns out to be wrong.

Kind of problem: Prediction

Recipe numbers to look at: 14

D

Industry: Production, especially consumer goods

Question: Can we predict a quality or safety problem from quality and service data at an early stage?

Target: Quality or safety issue arises (yes/no).

Input data: Data from customer service, returns and complaints, sales data for single items used to replace parts (implying that something has gone wrong with the product) and web and social media data if available. If transformation of data is necessary, the focus should be on preserving information about timing issues giving clues about when the product went wrong. Also, we have to consider the relationship between the number of problems and the quantity of sales of the original product, implying that we must preserve information about the defect rate. Also, any differences in the distribution of key input variables might be important, and we could consider including comparisons with previous periods or with other comparable products as separate additional input variables.

Cut-off: The cut-off depends on the cost of a decision that turns out to be wrong.

Kind of problem: Prediction

Recipe numbers to look at: 1 and 2

E

Industry: Medicine

Question: Is it possible to recognise a pattern related to a certain medical condition by analysing an image (e.g. from a microscope) automatically?

Target: The image shows the pattern according to expert opinion (yes/no). The difficulty is to summarise the expert opinion. It may happen that two or more experts come to different conclusions. So, a central concern is how to handle this uncertainty and decide whether the image has the pattern or not.

Input data: Data obtained from images with and without the pattern. The data should contain information like number and shapes of artefacts, description of shape lines (smooth, serrated, etc.), levels of grey and light/dark levels. If available, more personal data on individual patient history is also helpful.

Cut-off: The cut-off depends on the cost of a decision that turns out to be wrong.

Kind of problem: Prediction

Recipe numbers to look at: 1

F

Industry: Medicine

Question: Is it possible to predict the survival probability of seriously ill people in intensive care if they have to be removed because of shortage of places?

Target: Person will die quite soon after being removed (yes/no) according to expert opinion.

Input data: Personal medical history

Cut-off: This issue should be taken seriously. It may be better to have three classes in the target variable instead of two, for example, will be alive on 14th day after leaving intensive care, dies before the 14th day after leaving intensive care and unsure.

Kind of problem: Prediction

Recipe numbers to look at: 1 and 2

G

Industry: Medicine and production

Question: Is it possible to predict the contamination of an area (e.g. with a specific fungus)?

Target: Measured contamination (e.g. amount of fungus in a specific time period and location). The target can be continuous, classified or binary $(0 = OK, 1 = critical value reached)$.

Input data: All process data

Cut-off: Depends on the level of the target

Kind of problem: Prediction and description

Recipe numbers to look at: 1, 2, 10 and 14

H

Industry: Finance, risk and control

Question: Is it possible to predict the likelihood of repaying a loan or not?

Target: A binary variable: $0 =$ repay the loan and $1 =$ not to repay the loan

Input data: All data available to describe the clients. If they are quite new clients with little data traceability in your system, try to get additional data on the social and economic situation in the client's neighbourhood. Sometimes, publicly accessible data on creditworthiness is available and can be used either as a prediction in its own right or as an input variable in the prediction.

Cut-off: The cut-off depends on the cost of a decision that turns out to be wrong.

Kind of problem: Prediction

Recipe numbers to look at: 1 and 2

I

Industry: Finance, risk and control

Question: Is it possible to group clients into groups with homogenous needs?

Target: No, as this is an unsupervised learning problem

Input data: All data available to describe the clients. If they are quite new clients with little data traceability in your system, try to get additional data on the social and economic situation in the client's neighbourhood. Sometimes, publicly accessible data on creditworthiness is also available and can be used as an input variable in the analysis.

Cut-off: No

Kind of problem: Description and cluster analysis

Recipe numbers to look at: 13

J

Industry: Finance, risk and control

Question: Is it possible to recognise unusual patterns in a client account quite early?

Target: No. The problem can be solved with descriptive and exploratory statistics just using means, medians and variances.

Input data: All data available to describe the clients. If they are quite new clients with little data traceability in your system, try to get additional data on the social and economic situation in the client's neighbourhood. Sometimes, publicly accessible data on creditworthiness is also available and can be used either as a prediction in its own right or as an input variable.

Cut-off: No

Kind of problem: Description

Recipe numbers to look at: 10

These nine examples aim to show how to map your individual business problems to those described in the aforementioned chapters. However, data obtained in areas other than sales and marketing usually has much higher quality because most data is measured by sensors. Therefore, you should also consider using the well-established techniques of problem solving and data analysis associated with Lean, Six Sigma and Total Quality Management.

In our view 'big data' is less an analytical challenge, it is much more a challenge in Data Quality, Data Enrichment and Data Management. Data mapping and storage technologies are critical issues. Reading publications on 'big data', it is apparent that much of the analysis being carried out is based on ordinary statistical methods using well-known techniques such as SPC charts, and data mining; indeed, many of the solutions follow the methods given in our recipes. This shows the universal applicability of statistical and data mining techniques and should encourage new users to take up the challenge of working with data and try out the recipes described in this book.

13.5 FUTURE VIEWS

Analytics including data mining is becoming more and more important in all areas of everyday life. This is reflected in the increasing number of job opportunities for statisticians, data miners and data scientists. Web monitoring is changing the way goods and services are made available. In many aspects of our domestic and work life, the correct functioning of machines and equipment all depends on data mining techniques. As the world becomes increasingly automated, intelligent prediction systems are coming close to making human predictions obsolete. These powerful techniques can bring enormous benefits in terms of correctly pinpointing situations, people or opportunities. In all industry sectors, people expect that machines will become ever cleverer. However, machine learning is based on previous experience, and although this is a fundamental part of data mining, we must not undertake data mining in this purely mechanical way without adding our human ingenuity.

The danger with learning from past experience is that it can stifle innovation and curb imagination. If your search on the web is dominated by prediction models, then you have less chance to see new things; businesses can miss new opportunities because in the historic records they have no relevant data and no new patterns can be identified for the future. There are many nice examples where relying on past experience would inhibit originality. For example, if someone knows you like chocolates, you may end up with chocolates every year and never try anything new; if you just optimise the design of a horse-drawn coach, you will not discover an automobile; if you work on perfecting the design of a candle, you will not end up with a light bulb.

We have to keep the learning space open to expand and renew the data mining models. If there is no new input from outside, this cannot happen. We noted in the preceding text that it is sometimes useful to include some variables that have no proven relation to the problem at hand but are widely reported, as these variables can give an overview of the general environment and may become influential in the future. We also noted that one good reason for using a combination of methods when analysing data is to ensure that you do not lose the opportunity to observe the reaction of a group of people who are not considered important in the model but do exist in the population. You need to keep variety in the dataset so that when models are updated, there is a chance to pick up features that may have changed their importance.

The key message is to remain open to new methods and continue to collect a wide range of data and make sure that astonishing things can happen and new ideas can emerge. Powerful data mining techniques are making substantial changes to our lives. Provided we remain open to new ideas, we can enjoy the enormous benefits and look forward to continued discoveries.

Bibliography

Abraham, B. and Ledolter, J. (2005) Statistical Methods for Forecasting. Hoboken: Wiley-Interscience.

Aßmann, S. (2013) Social Media Monitoring, 1st edition http://social-media-monitoring. blogspot.de (accessed 4 January 2014)

Aerni, M. and Bruhn, M. (2008) Integrierte Kommunikation. Zürich: Compendio Bildungsmedien.

Agrawal, R., Imielinski, T. and Swami, A. N. (1993) Mining association rules between sets of items in large databases. In Proceedings of the 1993 ACM SIGMOD International Conference on Management of Data. Washington, DC: ACM Press.

Ahlemeyer-Stubbe, A. and Madsen, B. (2006) Data mining: which model comes out with the best model? A confrontation between regressions, decision trees and neural networks. Presented at the ENBIS Conference 2006, Wroclaw.

Albers, S. and Herrmann, A. (2002) Handbuch Produktmanagement. Strategieentwicklung, Produktplanung, Organisation, Kontrolle, 2nd revised and expanded edition. Wiesbaden: Gabler.

Alby, T. (2008) Web 2.0. Konzepte, Anwendungen, Technologien, 3rd revised edition. München: Hanser.

Altmetrics: A Manifesto (2013) http://altmetrics.org/manifesto/ (accessed 16 November 2013).

Amthor, A. and Brommund, T. (2010) Mehr Erfolg durch Web Analytics. München: Hanser.

Anahory, S. and Murray, D. (1997) Data Warehouse, Planung, Implementierung und Administration. Bonn/Reading: Addison-Wesley-Longman.

Appelrath, H.-J. (ed.) (1991) Datenbanksysteme in Büro, Technik und Wissenschaft. Berlin: Springer.

A Practical Guide to Data Mining for Business and Industry, First Edition.
Andrea Ahlemeyer-Stubbe and Shirley Coleman.
© 2014 John Wiley & Sons, Ltd. Published 2014 by John Wiley & Sons, Ltd.
Companion website: www.wiley.com/go/data_mining

Arteaga, F. and Ferrer, A. (2002) Dealing with missing data in MSPC: several methods, different interpretations, some examples. *Journal of Chemometrics*, 16, 408–418.

Au, G. and Choi, I. (1999) Facilitating implementation of quality management through information technology. *Information and Management*, 36, 287–299.

Backhaus, K., Erichson, B. and Weiber, B. (2010) Fortgeschrittene Multivariate Analysemethoden: Eine anwendungsorientierte Einführung. Heidelberg: Springer Verlag GmbH.

Bagusat, A. and Hermanns, A. (2008) E-Marketing-Management. Grundlagen und Prozesse für Business-to-Consumer-Märkte, 1st edition. München: Vahlen.

Balakrishnan, N., Colton, T., Everitt, B., Piegorsch, W., Ruggeri, F. and Teugels J. (Editors-in-Chief) (Forthcoming) WileyStatsRef: Statistics Reference Online. Chichester: John Wiley & Sons, Ltd.

Barquin, R. C. and Edelstein, H. A. (eds.) (1997a) Building, Using and Managing the Data Warehouse. Upper Saddle River: Prentice Hall PTR.

Barquin, R. C. and Edelstein, H. A. (eds.) (1997b) Planning and Designing the Data Warehouse. Upper Saddle River: Prentice Hall PTR.

Bazan, J. G., Nguyen, H. S., Nguyen, S. H., Synak P. and Wróblewski J. (2000) Rough set algorithms in classification problem. In Polkowski, L., Tsumoto, S. and Lin T. (eds.) Rough Set Methods and Applications. Heidelberg: Physica-Verlag.

Bazan, J., Nguyen, H. S., Skowron, A. and Szczuka, M. (2003) A View on Rough Set Concept Approximations. In Wang, G., Liu, Q., Yao, Y. and Skowron A. (eds.) Proceedings of R.S.F.D.G.r.C. Chongqing: Springer.

Becker, P. (2012) Twitter to hit 500 million accounts by February. http://wallblog.co.uk/2012/01/16/twitter-to-hit-500-million-accounts-by-february/ (accessed on 16 November 2013).

Bellinger, G. (2004) Systems thinking, knowledge management – emerging perspectives. http://www.systems-thinking.org/kmgmt/kmgmt.htm (accessed on 16 November 2013).

Berger, J., Betrò, B., Moreno, E., Pericchi, L. R., Ruggeri, F., Salinetti, G. and Wasserman, L. (eds.) (1996) Bayesian Robustness (Lecture Notes IMS, vol. 29). Hayward: Institute of Mathematical Statistics.

Berry, M. J. A. and Linoff, G. S. (2000) Mastering Data Mining. New York: John Wiley & Sons, Inc.

Berry, M. and Linoff, G. (2004) Data Mining Techniques, 2nd edition. Indianapolis: Wiley Publications, Inc.

Bose, I. and Mahapatra, K. R. (2001) Business data mining – a machine learning perspective. *Information and Management*, 39(3), 211–225.

Brauckmann, P. (ed.) (2010) Web-Monitoring, Gewinnung und Analyse von Daten über das Kommunikationsverhalten im Internet. Konstanz: UVK Verlagsgesellschaft mbH Broschiert.

Breitenstein, R. (2002) Memetik und Ökonomie. Wie die Information die Wirtschaft nach ihrem Interesse lenkt. http://www.heise.de/tp/artikel/13/13649/1 (accessed on 16 November 2013).

Breyfogle, F. W. (1999) Implementing Six Sigma. New York: John Wiley & Sons, Inc.

Bryman, A. and Bell, E. (2011) Business Research Methods, 3rd edition. Oxford: Oxford University Press.

Bunjes, M. (2012) Falle Internet: Gekaufte Freunde, bezahlte Tipps. http://www2.evan-gelisch.de/themen/medien/falle-internet-gekaufte-freunde-bezahlte-tipps56819 (accessed 25 September 2013). Frankfurt: Gemeinschaftswerks der Evangelischen Publizistik GmbH.

Buttle, F. (2009) Customer Relationship Management. Concepts and Technologies, 2nd edition. Oxford: Butterworth-Heinemann.

Caulcutt, R. (1995) Achieving Quality Improvement: A Practical Guide. London/New York: Chapman & Hall.

Caulcutt, R. (2008) Statisticians in industry in Coleman, S. Y., Greenfield, T., Stewardson, D. J. and Montgomery, D. (eds.) (2008) Statistical Practice in Business and Industry. Chichester: John Wiley & Sons, Ltd.

Cendrowska, J. (1987) PRISM: an algorithm for inducing modular rules. *International Journal of Man–machine Studies*, 27(4), 349–370.

Chamoni, P. and Gluchowski, P. (eds.) (1998) Analytische Informationssysteme: Data Warehouse, On-Line Analytical Processing, Data Mining. Berlin: Springer.

Chatfield, C. (1995) Problem Solving: A Statistician's Guide, 2nd edition. Boca Raton: Chapman & Hall/CRC Press.

Chatfield, C. (1998) Statistics for Technology. London: Chapman & Hall/CRC Press.

Chatfield, C. (2000) Time-Series Forecasting. London: Chapman & Hall/CRC Press.

Chatfield, C. (2004) The Analysis of Time Series: An Introduction, 6th edition. Boca Raton: CRC Press.

Chatfield, C. and Collins, A. (2000) Introduction to Multivariate Analysis. London: Chapman & Hall/CRC Press.

Codd, E. F. (1970) A relational model for large shared data banks. *Communications of the ACM*, 13(6), 377–387.

Codd, E. F. (1990) The Relational Model for Database Management: Version 2. Reading: Addison-Wesley.

Codd, S. B. and Salley, C. T. (1993) Providing OLAP (On-Line Analytical Processing) to User-Analysts: An IT Mandate. White Paper. San Jose: E. F. Codd & Associates.

Coleman, S. and Smith, K. (2007) Data mining sales data for Kansei Engineering. In Pham, D. T., Eldukhri, E. E. and Soroka, A. J. (eds.) Innovative Production Machines and Systems, 3rd IPROMS Virtual Conference. Boca Raton: CRC Press, pp. 268–273.

Coleman, S. Y., Greenfield, T., Stewardson, D. J. and Montgomery, D. (eds.) (2008) Statistical Practice in Business and Industry. Chichester: John Wiley & Sons, Ltd.

Corey, M. J. and Abbey, M. (1997) Oracle Data Warehousing. Berkeley: Osborne McGraw-Hill.

Cox, I. (2010) Visual Six Sigma – Making Data Analysis Lean. Hoboken: John Wiley & Sons, Inc.

Cunha C. D. A., Agard, E. and Kusiak, A. (2006) Data mining for improvement of product quality. *International Journal of Production Research*, 44, 4027–4041.

Cyganiak, R. and Jentzsch, A. (2011) The linking open data cloud diagram. http://richard.cyganiak.de/2007/10/lod/ (accessed on 16 November 2013).

Dahlgaard, J. J., Kristensen, K. and Kanji, K. (1998) Fundamentals of Total Quality Management. London: Chapman & Hall.

Dale, G. D. (1994) Managing Quality, 2nd edition. London: Prentice Hall.

Dawkins, R. (2007) Das egoistische Gen. München: Jubiläumsausgabe.

Duermyer, R. (ND) Viral marketing – internet viral marketing. http://homebusiness.about.com/od/homebusinessglossar1/g/viral-marketing.htm (accessed on 16 November 2013).

Ehrlenspiel, K. (2009) Integrierte Produktentwicklung – Denkabläufe, Methodeneinsatz, Zusammenarbeit, 4th edition. München: Hanser.

Eicker, S. and Schüngel, M. (1998) Stand der Unternehmensdaten Modellierung in der Praxis. *Information Management & Consulting*, 13, 78–85.

Evan, D. S. (2009) The online advertising industry: economics, evolution, and privacy. *The Journal of Economic Perspectives*, 23(3, Summer), 37–60(24). American Economic Association

Evandt, O., Coleman, S. Y., Ramalhoto, M. F. and van Lottum, C. (2004) A little-known robust estimator of the correlation coefficient and its use in a robust graphical test for bivariate normality with applications in the aluminium industry. *Quality and Reliability Engineering International*, 20(5), 433–456.

Faltin, F., Kenett, R. and Ruggeri, F. (eds.) (2012) Statistical Methods in Healthcare. Chichester: John Wiley & Sons, Ltd.

Fayyad, U., Piatetsky-Shapiro, G., Smyth, P. and Ramasami, U. (1996) Advances in Knowledge Discovery and Data Mining. Cambridge: MIT Press.

Fernandez, G. (2003) Data Mining Using SAS Applications. Boca Raton: Chapman & Hall/CRC Press.

Flach, P. A. and Lachiche, N. (1999a) Confirmation-guided discovery of first-order rules with Tertius. *Machine Learning*, 42(1–2), 61–95. Boston: Kluwer Academic Publishers.

Flach, P. A. and Lachiche, N. (1999b) The Tertius system. http://www.cs.bris.ac.uk/Research/MachineLearning/Tertius/ (accessed on 16 November 2013).

Fleisch, E. and Dierkes, M. (2003) Ubiquitous Computing aus betriebswirtschaftlicher Sicht. *Wirtschaftsinformatik*, 45(6), S611–S620.

Floemer, A. (2012) YouTube: Google nennt neue, beeindruckende Zahlen. http://t3n.de/news/youtube-google-nennt-neue-361320/ (accessed on 16 November 2013).

Frosch-Wilke, D. and Raith, C. (2002) Marketing-Kommunikation im Internet. Theorie, Methoden und Praxisbeispiele vom One-to-One- bis zum Viral-Marketing, 1st edition. Braunschweig: Vieweg.

Gabler Wirtschaftslexikon. Web 2.0. http://wirtschaftslexikon.gabler.de (accessed on 16 November 2013).

Gabriel, R. and Röhrs, H.-P. (1995) Datenbanksysteme: Konzeptionelle Datenmodellie-rung und Datenbankarchitekturen, 2nd edition. Berlin: Springer.

Gaßmair, D. (2009) Die Wahrheit über Virales Marketing. http://www.viralandbuzzmarketing.de/die-wahrheit-ueber-virales-marketing/ (accessed on 16 November 2013).

Gladwell, M. (2002) The Tipping Point. How Little Things Can Make a Big Difference, 1st edition. Boston: Back Bay Books.

Gluchowski, P. (1997) Data warehouse. *Informatik Spektrum*, 20(1), 48–49. Heidelberg: Springer Verlag GmbH.

Gluchowski, P., Gabriel, R. and Chamoni, P. (1997) Management Support Systeme. Computergestützt Informationssysteme für Führungskräfte und Entscheidungsträger. Berlin: Springer.

Groth, R. (1999) Data Mining: Building Competitive Advantage. Upper Saddle River: Prentice Hall.

Habermas, J. (2001) Die Zukunft der menschlichen Natur. Auf dem Weg zu einer liberalen Eugenik?, 1st edition. Frankfurt am Main: Suhrkamp.

Hague, P. (2002) Market Research. A Guide to Planning, Methodology and Evaluation, 3rd edition. London: Kogan Page.

Han, J., Kamber, M. and Pei, J. (2011) Data Mining: Concepts and Techniques, 3rd edition. San Francisco: Morgan Kaufmann (Previous editions by J. Han and M. Kamber, 2000, 2006).

Hand, D. J., Mannila, H. and Smyth, P (2001) Principles of Data Mining. New York: MIT Press.

Hartung, B. (2012) Social Media. Nutzerzahlen im Januar 2012. http://birgerh. de/2012/02/03/social-media-nutzerzahlen-im-januar-2012/ (accessed on 16 November 2013).

Hartung, J., Elpelt, B. and Klösener, K. (2005) Statistik: Lehr- und Handbuch der angewandten Statistik, 14th edition. Oldenbourg: München Wien.

Hartung, J., Knapp, G. and Sinha, B. K. (2011) Statistical Meta-Analysis with Applications (Wiley Series in Probability and Statistics). Hoboken: John Wiley & Sons, Inc.

Heller, C. (2009) Klartext: was ist ein Meme? http://www.netzpiloten.de/klartext-was-ist-eine meme/ (accessed 25 September 2013), Netzpiloten AG, Hamburg.

Henderson, G. R. (2006) Six Sigma Quality Improvement with MINITAB. Hoboken: John Wiley & Sons, Inc.

Holte R. C. (1993) Very simple classification rules perform well on most commonly used datasets. *Machine Learning*, 11, 63–90. Boston: Kluwer Academic Publishers.

Homburg, C. and Krohmer, H. (2006a) Marketingmanagement. Strategie, Instrumente, Umsetzung. Wiesbaden: Gabler.

Homburg, C. and Krohmer, H. (2006b) Marketingmanagement. Studienausgabe: Strategie, Instrumente, Umsetzung, Unternehmensführung, 2nd edition. Wiesbaden: Gabler.

Hotz, A., Halbach, J. and Schleinhege, M. (2010) Social Media im Handel, Ein Leitfaden für kleine und mittlere Unternehmen, 1st edition. Köln (eds.): E-Commerce-Center Handel; Hamburg: Clever and Smart Public Relations.

Hughes, A. M. (2003) The Customer Loyalty Solution. New York: McGraw-Hill Professional.

Hughes, A. M. (2005) Strategic Database Marketing. New York: McGraw-Hill Professional.

Inmon, W. H. (1996) Building the Data Warehouse, 2nd edition. New York: John Wiley & Sons, Inc.

Inmon, W. H. and Hackathorn, R. D. (1994) Using the Data Warehouse. New York: John Wiley & Sons, Inc.

Jefkins, F. (1998) Public Relations, 5th edition. London: Financial Times.

Jütte, W. (2002) Soziales Netzwerk Weiterbildung. Analyse lokaler Institutionslandschaften. http://www.die-bonn.de/doks/juette0201 (accessed on 16 November 2013).

Kanji, K. G. and Asher, M. (1996) 100 Methods for Total Quality Management. London: Sage Publications.

Kantrardzic, M. (2003) Data Mining: Concepts, Models, Methods, and Algorithms. Hoboken: IEEE Press.

Kasper, H., Dausinger, M., Kett, H. and Renner, T. (2010) Fraunhofer IAO, Marktstudie Social Media Monitoring Tools. ITLösungen zur Beobachtung und Analyse unternehmensstrategisch relevanter Informationen im Internet, 1st edition. Stuttgart: Fraunhofer-Institut für Arbeitswirtschaft und Organisation.

Kaushik, A. (2007) Web Analytics an Hour a Day. Hoboken: John Wiley & Sons, Inc.

KDnuggets. http://www.kdnuggets.com/ (accessed on 16 November 2013).

Kenett, R. and Zacks, S. (1998) Modern Industrial Statistics: Design and Control of Quality and Reliability. Pacific Grove: Duxbury/Wadsworth Publishing.

Kenett, R. and Raanan, Y. (eds.) (2010) Operational Risk Management: A Practical Approach to Intelligent Data Analysis. Chichester: John Wiley & Sons, Ltd., http://eu.wiley.com/WileyCDA/WileyTitle/productCd-047074748X.html (accessed on 16 November 2013).

Kenett, R. S. and Salini, S. (2011) Modern analysis of customer satisfaction surveys: comparison of models and integrated analysis. *Applied Stochastic Models in Business and Industry*, 27, 465–475.

Kenett, R. S. and Shmueli, G. (2013) On information quality. *Journal of the Royal Statistical Society*. doi: 10.1111/rssa.12007.

Kenett, R. S., Coleman, S. Y. and Stewardson, D. J. (2003) Statistical efficiency – the practical perspective. *Quality and Reliability Engineering International*, 19, 265–272.

Klau, P. (2009) So funktioniert Twitter. Eine Kurzanleitung zum Zwitschern im Web. http://peter-klau.suite101.de/so-funktioniert-twitter-a57535 (accessed on 16 November 2013).

Kolarik, J. W. (1995) Creating Quality: Concepts, Systems, Strategies and Tools. New York: McGraw-Hill.

Kortmann, C. (2008) Virales Marketing auf YouTube. Die gekaufte Weisheit der Vielen. http://www.sueddeutsche.de/kultur/virales-marketing-auf-youtube-die-gekaufte-weisheit-der-vielen-1.587853 (accessed on 26 September 2013).

Kotler, P. (2007) Grundlagen des Marketing, 4th updated edition. München: Pearson Studium.

Kotler, P., Keller, K. and Bliemel, F. (2007) Marketing-Management, Strategien für wertschaffendes Handeln, 12th edition. München: Pearson Studium.

Kozinets, R. (2009) Netnography: Doing Ethnographic Research Online. Los Angeles/ London: Sage Publications.

Kum, H. C., Chang, J. H. and Wang, W. (2007) Benchmarking the effectiveness of sequential pattern mining methods. *Data & Knowledge Engineering*, 60(1), 30–50.

Kumar, V. and Petersen, J. A (2012) Statistical Methods in Customer Relationship Management. Chichester: John Wiley & Sons, Ltd.

Langner, S. (2005) Viral Marketing: Wie Sie Mundpropaganda gezielt auslösen und Gewinn bringend nutzen, 1., Auflage. Wiesbaden: Gabler.

Laningham, S. (2006) developerWorks Interview: Tim Berners-Lee. http://www. ibm.com/developerworks/podcast/dwi/cm-int082206txt.html (accessed on 16 November 2013).

Larsen, B. S. and Madsen, B. (1999) Error identification and imputations with neural networks. Paper presented at the UN/ECE work session on statistical data editing, Rome.

Ledolter, J. and Swersey, A. (2007) Testing 1-2-3. Stanford: Business Books.

Lehner, F. and Maier, R. (1994) Information in Betriebswirtschaftslehre, Informatik und wirt-schaftsinformatik, Forschungsbericht Nr.1. der Schriftenreihe des Lehrstuhls für wirtschaftsin-formatik und Informationsmanagement, Wissenschaftliche Hochschule für Unternehmensfüh-rung, Koblenz.

Linacre, J. M. (1999) Understanding Rasch measurement: estimation methods for Rasch measures. *Journal of Outcome Measurement*, 3(4), 382–405.

Lindsay, M. W. and Petrick, A. J. (1997) Total Quality and Organization Development. Delray Beach: St. Lucie Press.

Link, J. (1997) Handbuch des Database Marketing. Ettlingen: IM-Fachverl. Marketing-Forum.

Linoff, G. and Berry, M. (2011) Data Mining Techniques: For Marketing, Sales, and Customer Relationship Management, 3rd edition. Indianapolis: Wiley Publications, Inc.

Locke, C., Searls, D., Weinberger, D. and Levine, R. (1999) The Cluetrain Manifesto. http://www.cluetrain.com (accessed on 16 November 2013).

Loveman, G. (2003) Diamonds in the data mine. *Harvard Business Review* (May), 109–123.

Maaß, C. (2007) ZP-Stichwort: Semantisches Web. *Zeitschrift für Planung & Unternehmenssteuerung*, 18(1), S123–S129.

Madsen, B. (2011) Statistics for Non-Statisticians. New York: Springer.

Martin, W. (ed.) (1998) Data Warehousing. Bonn: International Thomson Publishing GmbH.

Maurice, F. (2007) Web 2.0 Praxis. AJAX, Newsfeeds, Blogs, Microformats, 1., Auflage. München: Markt + Technik.

McCollin, C. and Coleman, S. Y. (2013) Historical published maintenance data: what can it tell us about reliability modelling? *Quality and Reliability Engineering International*. doi: 10.1002/qre.1585.

McCullagh, P. (1980) Regression models for ordinal data. *Journal of the Royal Statistical Society*, 42(2), 109–142.

McCullagh, P. and Nelder, J. A. (1989) Generalised Linear Models, 2nd edition. London: Chapman & Hall.

Meffert, H., Burmann, C. H. and Kirchgeorg, M. (2008) Marketing, Grundlagen marktorientierter Unternehmensführung Konzepte, Instrumente, Praxisbeispiele, 10th edition. Wiesbaden: Gabler.

Mitchell, T. (1997) Machine Learning. New York: McGraw-Hill.

Mitchel, T. M. (2006) The Discipline of Machine Learning. http://www.cs.cmu.edu/~tom/pubs/MachineLearning.pdf (accessed on 16 November 2013).

Mizuno, S. (1988) Management for Quality Improvement: The Seven New QC Tools. Cambridge: Productivity Press.

Monness, E. and Coleman, S. Y. (2006) LISREL: an alternative to MANOVA and principal components in designed experiments when the response is multidimensional. *Quality and Reliability Engineering International*, 22(2), 213–224.

Montgomery, D. C. (2008) Design and Analysis of Experiments. Hoboken: John Wiley & Sons, Inc.

Mucksch, H. and Behme, W. (eds.) (1998) das Data Warehouse-Konzept. Architektur-Datenmodelle-Anwendungen, 3rd edition. Wiesbaden: Gabler.

Müller, J. (2000) Transformation operativer Daten zur Nutzung im Data Warehouse. Wiesbaden: Deutscher Universitäts-Verlag/Gabler.

Münker, S. (2009) Die sozialen Medien des Web 2.0. In Michelis, D. and Schildhauer, T. (eds.) Social-Media-Handbuch. Theorien, Methoden, Modelle, 1st edition. Baden-Baden: Nomos, pp. S31–S42.

Niederhuber, K. (2011) Die Komposition macht den Unterschied. http://corporateaudioblog.twoday.net/ (accessed on 16 November 2013).

Nielsen (2009) Nielsen global online consumer survey – trust, value and engagement in advertising. http://de.nielsen.com/pubs/documents/NielsenTrustAdvertising GlobalReport July09.pdf (accessed on 16 November 2013).

Nielsen (2011) State of the media – the social media report Q3 2011. http://www.nielsen.com/content/dam/corporate/us/en/reports-downloads/2011-Reports/nielsen-social-media-report.pdf (accessed on 16 November 2013).

Nieschlag, R., Dichtl, E. and Hörschgen, H. (2002) Marketing, 19th revised and expanded edition. Berlin: Duncker & Humblot Verlag.

Nordbotten, S. (1995) Editing statistical records by neural networks. *Journal of Official Statistics*, 11(4), 391–411.

Oetting, M. (2006) Wie das Web 2.0 das Marketing revolutioniert. In Schwarz, T. (ed.) Leitfaden integrierte Kommunikation, 1st edition. Waghäusel: Absolit, Dr. Schwarz Consulting, pp. S173–S195.

OLAP Council (1995) OLAP and OLAP server definitions, The OLAP council, 1995. http://www.olapcouncil.org (accessed on 16 November 2013).

O'Reilly, T. (2005) What is Web 2.0. Design patterns and business models for the next generation of software. http://oreilly.com/web2/archive/what-is-web-20.html (accessed on 16 November 2013).

Parr-Rudd, O. (2000) Data Mining Cookbook: Modeling Data for Marketing, Risk, and Customer Relationship Management. New York: John Wiley & Sons, Inc.

Perner, P. (ed.) (2003) Advances in Data Mining. New York: Springer.

Perner, P. (ed.) (2006) Advances in Data Mining. New York: Springer.

Perner, P. (ed.) (2008) Case-Based Reasoning and the Statistical Challenges. Berlin/New York: Springer.

Perreault, W. D. and McCarthy, J. (1996) Basic Marketing. New York: McGraw-Hill.

Petereit, D. (2011) Twitter verdreifacht Anmeldezahlen seit iOS5-Start. http://t3n.de/news/twitter-verdreifacht-anmeldezahlen-seit-ios5-start-337461/ (accessed on 16 November 2013).

Piatetsky-Shapiro, G., Frawley, W. J. and Matheus, C. (1991) Knowledge Discovery in Databases. Menlo Park: A.A.A.I./MIT Press.

Poessneck, L. (2008) Web 2.0 ist erst der Anfang. Interview mit Wolfgang Wahlster. http://www.silicon.de/39192819/web-2-0-ist-erst-der-anfang/ (accessed on 16 November 2013).

Quinlan, J. R. (1986) Induction of decision trees. *Machine Learning*, 1(1), 81–106.

Quinlan, J. R. (1992) C4.5: Program for Machine Learning. San Mateo: Morgan Kaufmann.

Refaat, M. (2007) Data Preparation for Data Mining Using SAS. Amsterdam/Boston: Morgan Kaufmann.

Reif, G. (2006) Semantische Annotation. Semantic Web. In Pelligrini, T. and Blumauer, A. (eds.) Semantic Web. Wege zur vernetzten Wissensgesellschaft. Heidelberg: Springer, pp. S405–S418.

Renker, L. C. (2008) Virales Marketing im Web 2.0. Innovative Ansätze einer interaktiven Kommunikation mit dem Konsumenten, 1st edition. München: IFME.

Rexer Analytics (2011) Data miner survey. http://www.rexeranalytics.com/index.html (accessed on 16 November 2013).

Rios Insua, D. and Ruggeri, F. (2000) Robust Bayesian Analysis (Lecture Notes in Statistics). New York: Springer.

Rios Insua, D., Ruggeri, F. and Wiper, M. P. (2012) Bayesian Analysis of Stochastic Process Models. Chichester: John Wiley & Sons, Ltd.

Ripley, B. D. (2007) Pattern Recognition and Neural Networks. Cambridge/New York: Cambridge University Press.

Rogers, E. M. (2003) Diffusion of Innovations, 5th edition. New York: Free Press.

Röttger, U. (2000) Public Relations – Profession und Organisation – Öffentlichkeitsarbeit als Organisationsfunktion. Eine Berufsfeldstudie, 1st edition. Wiesbaden: VS Verlag.

Ruggeri, F., Kenett, R. and Faltin, F. (eds.) (2007) Encyclopedia of Statistics in Quality and Reliability. Chichester: John Wiley & Sons, Ltd.

Saritha, J. S., Govindarajulu, P., Prasad, R. K., Ramana Rao, S. C. V. and Lakshmi C. (2010) Clustering methods for credit card using Bayesian rules based on K-means classification. *International Journal of Advanced Computer Science and Applications*. 1(4), 92–95.

Scheer, A.-W. (1988) Wirtschaftsinformatik: Informationssysteme im Industriebetrieb. Berlin: Springer.

Schmalen, H. and Xander, H. (2002) Produkteinführung und Diffusion. In Albers, S. and Hermann, A. (eds.) Handbuch Produktmanagement. Strategieentwicklung, Produktplanung, Organisation, Kontrolle, 2nd revised and expanded edition. Wiesbaden: Gabler.

Schnell, R., Hill, P. B. and Esser, E. (2005) Methoden der empirischen Sozialforschung, 7th revised and expanded edition. München: Oldenbourg Verlag.

Schüller, A. M. (2011) Zukunftstrend Empfehlungsmarketing. Der beste Umsatzbeschleuniger aller Zeiten, 5th revised edition. Göttingen: BusinessVillage.

Schulz, S. (2009) Wir werden Echtzeit-Marketing lernen – oder untergehen. http://www.spiegel.de/wirtschaft/unternehmen/0,1518,657867,00.html (accessed on 16 November 2013).

Schürg, R. (2008) Studie: Viral Marketing funktioniert nur crossmedial. http://lingner.com/zukunftskommunikation/studie-viralmarketing-funktioniert-nur-crossmedial (accessed on 16 November 2013).

Schwarz, T. (2007) Leitfaden Online Marketing – 28 innovative Praxisbeispiele. Waghäusel: Marketing-Börse.

Schwarz, T. (2008) Praxistipps Dialog Marketing Vom Mailing bis zum Online-Marketing. Waghäusel: Marketing-Börse.

SEMPO Institute Glossary. http://www.sempo.org/?page=glossary (accessed on 16 November 2013).

Smith, E. V., Jr. and Smith, R. M. (eds.) (2004) Introduction to Rasch Measurement Theory, Models and Applications. Maple Grove: JAM Press.

Steve Toms. http://www.stevetoms.net/glossary.htm (accessed on 16 November 2013).

Stone, B. (2004) Who Let the Blogs Out? A Hyperconnected Peek at the World of Weblogs, 1., Auflage. New York: St. Martin's Griffin.

Sung, H. H. and Sang, C. P. (2006) Service quality improvement through business process management based on data mining. *ACM SIGKDD Explorations Newsletter*, 8, 49–56.

Szugat, M., Lochmann, C. and Gewehr, E. J. (2006) Social Software. schnell + kompakt, 1st edition. Frankfurt am Main: Entwickler Press.

Tsiptsis, K. and Chorianopoulos, A. (2009) Data Mining Techniques in CRM. Chichester/West Sussex: John Wiley & Sons, Ltd.

Tsironis, L., Bilalis, N. and Moustakis, V. (2005) Using machine learning to support quality management: framework and experimental investigation. *The TQM Magazine*, 17, 237–248.

van Lottum, C., Pearce, K. and Coleman, S. (2006) Features of Kansei engineering characterizing its use in two studies: men's everyday footwear and historic footwear. *Quality and Reliability Engineering International*, 22(6), 629–650.

Van Someren, M. and Urbančič, T. (2006) Applications of machine learning: matching problems to tasks and methods. *The Knowledge Engineering Review*, 20, 363–402.

Walsh, G., Hass, B. and Kilian, T. (2011) Grundlagen des Web 2.0. In Walsh, G., Hass, B. and Kilian, T. (eds.) Web 2.0. Neue Perspektiven für Marketing und Medien, 2nd revised and expanded edition. Berlin: Springer.

Warner, B. and Misra, M. (1996) Understanding neural networks as statistical tools. *The American Statistician*, 50, 284–293.

Webster's New World College Dictionary. (1999) 4th edition. John Wiley & Sons.

Wheeler, D. J. (2002) Two plus two is only equal to four on the average. http://www.spcpress.com/ink_pdfs/wh_two_plus_two.htm (accessed on January 2007).

Witten, I. H., Frank, E. and Hall, M. A. (2011) Data Mining: Practical Machine Learning Tools and Techniques, 3rd edition. Burlington: Morgan Kaufmann. (Previous editions by I. H. Witten and E. Frank, 2000, 2005.)

Wittmann, W. (1959) Unternehmung und unvollkommene Information. Köln/Opladen: Westdeutscher Verlag.

Zerfass, A. and Sandhu, S. (2008) Interaktive Kommunikation, Social Web und Open Innovation: Herausforderungen und Wirkungen im Unternehmenskontext. Köln: Herbert von Halem Verlag.

Zideate (Marketing Dictionary). http://www.zideate.com/dictionary (accessed on 16 November 2013).

Index

Note: Page numbers in *italics* refer to Figures.

Printed and bound by CPI Group (UK) Ltd, Croydon, CR0 4YY

27/10/2024

14580190-0002